AF829322

'In this carefully researched book, Whittington invites the reader to explore how Indigenous cultural tourism may – or may not – change visitors' thinking, attitudes and behaviour. The author urges us to consider how our engagement with Indigenous culture and Country might become transformational rather than romantic, insisting we consider the complex entanglements of settler colonialism and the tourist experience. This is an important work that deserves to be read.'

Professor Sarah Maddison, *BA, PhD, MAIATSIS, Director, The (so-called) Australian Centre, School of Social and Political Sciences, The University of Melbourne, Australia*

'Whittington's sensitive and detailed empirical analysis of visitor responses to Indigenous cultural heritage at two World Heritage Sites shows us how the visitor experience at such places can both challenge and perpetuate colonial narratives. It provides a template for heritage policymakers, managers and tour operators on the forms of interpretation and experience which may be most efficacious in transforming visitor attitudes and promoting reconciliation in settler-colonial contexts, and for heritage researchers in its methods and approaches to the study of the affective experience of heritage landscapes for visitors.'

Professor Rodney Harrison, *Professor of Heritage Studies, University College London, United Kingdom*

'Travel beyond the familiarity of one's environs has long been regarded as one of the best educators. In linking the interpretation of Indigenous tourism to the transformation of the visitor, Whittington tackles the sticky challenge of how this might aid reconciliation between Indigenous hosts and non-Indigenous guests in a settler-colonial milieu. In broadly querying interpretation and the modes of interpretation that are optimal, this book draws from theoretical and socio-historical framing that elucidates the ripple effects of cross-cultural touristic encounters. Whether they lead to fertile exchanges or reinforce hackneyed tropes and power disparities is a central contention.'

Professor Joseph M. Cheer, *Professor of Sustainable Tourism and Heritage, Western Sydney University, Australia*

Visitors, Settler Colonialism and Aboriginal Cultural Heritage

This book analyses visitor responses to the interpretation of Aboriginal cultural heritage in Australian protected areas, focusing on Uluṟu-Kata Tjuṯa National Park in the Northern Territory and the Greater Blue Mountains World Heritage Area in New South Wales.

It provides insights into the discursive features that structure various forms of interpretation of Aboriginal cultural heritage at these locales, ranging from on-site interpretative signage to audio tours available on mobile phone applications. It draws on visitors' firsthand accounts of the experience of participating in Traditional Custodian led cultural tours and camps, and the visitor learnings that resulted from these. Based on extensive interviews with visitors, the author argues that visitor responses to these experiences both perpetuate and challenge settler-colonial assumptions about Aboriginal peoples and their cultures in both more urban and remote locations. The book provides insight into the forms of interpretation that foster visitor transformations, thereby advancing a politics of reconciliation between Indigenous and non-Indigenous Australians, and the types of interpretation that may hinder such transformations, by reinforcing settler-colonial discourses and affective states.

The book is aimed at students and academics attempting to develop a more critical practice in relation to heritage interpretation, tourism, Aboriginal and Torres Strait Islander cultural heritage and settler-colonialism. It will also appeal to heritage professionals, cultural tour operators and agencies responsible for the provision of protected area interpretation, including both government sector and Indigenous organisations.

Vanessa Whittington holds a Doctorate from the Institute for Culture and Society, Western Sydney University (WSU) and currently works as a Sessional Academic with the School of Social Sciences, WSU. Her Doctoral research was in the field of critical heritage and tourism studies. She has a Master's Degree in Museum and Heritage Studies, a Master of Arts (Hons) in Women's Studies/Interdisciplinary Studies and a Bachelor of Arts with majors in History and Political Science. She has previously worked in senior policy and research roles for government and non-government agencies with a human services, social justice and social policy focus. She has published in the areas of Indigenous-visitor relations in protected area landscapes; heritage and contested memory in a settler-colonial context; the marginalisation of working-class urban heritage and world heritage discourse and gender.

Critical Studies in Heritage, Emotion and Affect
Series Editors:
Divya P. Tolia-Kelly, *University of Sussex*
Emma Waterton, *University of York*

This book series, edited by Divya P. Tolia-Kelly and Emma Waterton, is dedicated to Professor Steve Watson. Steve was a pioneer in heritage studies and was inspirational in both our personal academic trajectories. We, as three editors of the series, started this journey together, but alas we lost his magnificent scholarship and valued counsel too soon.

The series brings together a variety of new approaches to heritage as a significant affective cultural experience. Collectively, the volumes in the series provide orientation and a voice for scholars who are making distinctive progress in a field that draws from a range of disciplines, including geography, history, cultural studies, archaeology, heritage studies, public history, tourism studies, sociology and anthropology – as evidenced in the disciplinary origins of contributors to current heritage debates. The series publishes a mix of speculative and research-informed monographs and edited collections that will shape the agenda for heritage research and debate. The series engages with the concept and practice of Heritage as co-constituted through emotion and affect. The series privileges the cultural politics of emotion and affect as key categories of heritage experience. These are the registers through which the authors in the series engage with theory, methods and innovations in scholarship in the sphere of heritage studies.

Unfolding Spatial Movements in the Second-Hand Book Market in Kolkata
Notes on the Margins in the *Boipara*
Diti Bhattacharya

Visitors, Settler Colonialism and Aboriginal Cultural Heritage
Australian Protected Areas as Transformative Landscapes
Vanessa Whittington

For more information about this series, please visit: www.routledge.com/Critical-Studies-in-Heritage-Emotion-and-Affect/book-series/CSHEA

Visitors, Settler Colonialism and Aboriginal Cultural Heritage

Australian Protected Areas as Transformative Landscapes

Vanessa Whittington

Routledge
Taylor & Francis Group
LONDON AND NEW YORK

First published 2025
by Routledge
4 Park Square, Milton Park, Abingdon, Oxon OX14 4RN

and by Routledge
605 Third Avenue, New York, NY 10158

Routledge is an imprint of the Taylor & Francis Group, an informa business

© 2025 Vanessa Whittington

The right of Vanessa Whittington to be identified as author of this work has been asserted in accordance with sections 77 and 78 of the Copyright, Designs and Patents Act 1988.

All rights reserved. No part of this book may be reprinted or reproduced or utilised in any form or by any electronic, mechanical, or other means, now known or hereafter invented, including photocopying and recording, or in any information storage or retrieval system, without permission in writing from the publishers.

Trademark notice: Product or corporate names may be trademarks or registered trademarks, and are used only for identification and explanation without intent to infringe.

British Library Cataloguing-in-Publication Data
A catalogue record for this book is available from the British Library

ISBN: 978-1-032-65495-9 (hbk)
ISBN: 978-1-032-65496-6 (pbk)
ISBN: 978-1-032-65497-3 (ebk)

DOI: 10.4324/9781032654973

Typeset in Sabon
by KnowledgeWorks Global Ltd.

Contents

Figures

Acknowledgements

I am deeply grateful to Anangu, Gundungurra and Dharug Traditional Custodians of Uluru-Kata Tjuta National Park and the Greater Blue Mountains World Heritage Area for allowing me to conduct research on their traditional lands, and to those Traditional Custodians who agreed to participate in my research. My thanks also go to all other research participants, whether they be visitors, cultural tour providers or staff of agencies responsible for managing protected areas, for their generosity in sharing their time with me. I am particularly grateful to Steve Baldwin from Parks Australia for the support and ongoing assistance he provided, and to Brad Moore, formerly of Blue Mountains City Council, for his advice and support. Thanks also go to the academic supervisors of the thesis on which this book is based, Professors Emma Waterton and Denis Byrne. I am appreciative of the financial support provided for the research by the Institute for Culture and Society, Western Sydney University, and the Australian Government, through an Australian Government Research Training Program Scholarship. Finally, I would like to thank my partner David Ambery, without whose love and support this book would not have been written.

Abbreviations

ANPSAB	American National Park Service Advisory Board
ANTAR	Australians for Native Title and Reconciliation
APY Lands	A<u>n</u>angu, Pitjantjatjara, Yankunytjatjara Lands
BMNP	Blue Mountains National Park
COVID-19	Coronavirus disease of 2019
EPBC Act	Environment Protection and Biodiversity Conservation Act 1999 (C'th)
GBMWHA	Greater Blue Mountains World Heritage Area
ILUA	Indigenous Land Use Agreement
IPA	Indigenous Protected Area
IUCN	International Union for Conservation of Nature
NRM	Natural Resource Management
NSW	New South Wales
NSW NPWS	New South Wales National Parks and Wildlife Service
NT	Northern Territory
UKTNP	Ulu<u>r</u>u-Kata Tju<u>t</u>a National Park
UNESCO	United National Educational, Scientific and Cultural Organisation

1 Introduction

This book investigates the question of how visitors respond to the interpretation of Aboriginal cultural heritage in Australian protected areas,[1] with a view to learning whether exposure to such interpretation deepens visitors' understandings of Aboriginal peoples and cultures.[2] This has a bearing on the broader question of the potential for Aboriginal cultural tourism to engender forms of "transformational" change in the non-Aboriginal visitor to further an agenda of social justice and reconciliation between Aboriginal and non-Aboriginal Australians. Reconciliation has been defined as a "process where Aboriginal and Torres Strait Islander peoples, non-Indigenous Australians and Australian governments forge a new relationship based on mutual understanding, recognition and respect" (ANTAR n.d). I understand reconciliation to be a first step towards decolonisation in the Australian context, rather than an endpoint, and am mindful of the critique that reconciliation can be overly focused on fostering national unity and harmonious race relations at the expense of transformational social change to address issues of social justice for Aboriginal and Torres Strait Islander peoples (Little and McMillan 2017).

My research responds to a growing call to decolonise heritage and tourism scholarship and practice (Knudsen and Andersen 2019; Knudsen et al. 2022; Kølvraa and Knudsen 2020; Walter 2021; Wijesinghe et al. 2019), and addresses a gap in research on decolonising heritage within settler-colonial nations such as Australia, which were the former colonies of European colonial powers. Influenced by decolonial theorists such as Patrick Wolfe (1999, 2), I start from the premise that Australia is a settler-colonial nation where the colonisers came "to stay", making "invasion … a structure not an event". The resulting social and political system is reflected not only in formal government structures but also encompasses socio-cultural practices such as heritage and tourism, and the apparently neutral protected areas management practices of state agencies, increasingly recognised as laden with Eurocentric ontological assumptions (Howitt et al. 2013; Muller et al. 2019; Searle and Muller 2019). In this context, I understand state-sponsored heritage interpretation as not simply a method of communication targeting

DOI: 10.4324/9781032654973-1

the visitor that fosters objective learning but as a system of representation operating in the context of settler-colonial Australia that may contain a range of "postcolonising" (Moreton-Robinson 2015, 10) assumptions with material effects. I use the term "postcolonising" after Goenpul scholar Aileen Moreton-Robinson (2015, 10), rather than the term "postcolonial", to recognise that "[i]n Australia the colonials did not go home", and that colonisation is, in effect, continuing (see Chapter 2).

Although my focus is on decolonising heritage and tourism practice and scholarship, my understanding of what this entails is informed by the work of critical Indigenous scholars such as Linda Tuhiwai-Smith (2021), Sherry Copenace et al. (2022) and Eve Tuck and K. Wayne Yang (2012) on decolonisation more broadly, including research methodologies and educational institutions. There is a general consensus in this scholarship that decolonisation does not simply involve the Indigenisation of institutions or methodologies through the inclusion of Indigenous perspectives within existing systems, it also entails a challenge to the definition of what constitutes knowledge and who gets to decide this. Relatedly, decolonial scholars, such as those discussed in Chapter 2, have articulated a sustained critique of knowledge production rooted in the minority world, based on an Enlightenment view of knowledge arising in Eurocentric ontologies and epistemologies (Mignolo 2011; Quijano 2007).

In taking a critical approach to heritage interpretation in protected areas, I align myself with scholars who argue that the field of interpretation is under-theorised, market-driven and is unreflective of the theoretical underpinnings and values of its own practice (Ablett and Dyer 2010; Finegan 2019; Gonzalez and Vazquez 2014; Staiff 2014; Staiff et al. 2002). This lack of critical awareness is apparent in what I identify as a nature/culture dualism within protected area interpretation that positions Indigenous peoples as simply another part of the "natural" world (Langton 1996; Robin 2014). Other forms of misrepresentation of Indigenous peoples within protected area heritage interpretation identified in the limited existing literature include the presentation of Indigenous cultures as of the past and the relationships of Indigenous peoples with these landscapes as ending with colonisation (Ballantyne 1995; Carr 2004; Clarke and Waterton 2015; Porter 2010; Staiff et al. 2002). I also argue that there are gaps and silences in state-sponsored interpretive narratives, including an aversion to addressing contentious issues of historical injustice (Ballantyne 1995; Ballantyne et al. 2012; Clarke and Waterton 2015). Furthermore, there is limited research that offers the perspectives of Indigenous peoples on the interpretation of their own cultural heritage in protected areas or on tourism to places of importance to them, per se (Finegan 2019; Ruhanen and Whitford 2019; Whitford and Ruhanen 2016; Wijesinghe et al. 2019, but see, for example, Beck and Somerville [2002] and Travesi [2018]).

My research addresses these gaps by consulting with the Traditional Custodians of protected areas, including Aboriginal cultural tour providers, to obtain their perspectives on how they want their cultures interpreted to visitors, if at all. It applies a Foucauldian discourse analysis to state-sponsored protected area interpretation, to reveal discursive structures that authorise certain representations of Aboriginal cultural heritage and prohibit others (Fforde et al. 2013; Foucault 1981, 2002; Jäger and Maier 2016). Comparing state-sponsored forms of interpretation in protected areas, such as interpretive signage and ranger guided walks, with Traditional Custodian led cultural tours, this book builds upon existing research in the field of Indigenous cultural tourism (Bawaka Country et al. 2013; Palomino-Schalscha 2018; Ruhanen and Whitford 2019; Travesi 2018) to take up the important question of how Aboriginal control of the interpretive experience influences the types of messages conveyed, and visitor responses to these. I do this not only through discourse analysis but also by means of ethnographic methods such as visitor interviews, participant observation of cultural tours led by Aboriginal peoples, and a practice of auto-ethnography in which I record and reflect on my own responses to these experiences as a participant observer, both textually and visually (Pink 2015), through the artwork included herein. The use of auto-ethnography positions me as part of the research process, acknowledging the way that my standpoint has influenced the research, and reflects an openness to exploring how I may have been changed by it (Koopman et al. 2020).

My use of methodologies for the study of affect, such as visitor interviewing, also arises from an awareness that visitors are not passive objects who simply absorb interpretive messages. Pedagogical theory focusing on the way people learn and empirical heritage and tourism research investigating visitor expectations, reasons for visiting and the types of experience valued, demonstrates visitors' active role in meaning-making, and that affect, embodiment and emotion play a role in this process (Fraser 2007; Watkins et al. 2015; Witcomb 2013, 2015a, b). Influenced by developments within phenomenology and contemporary post-human philosophy, I recognise that ways of knowing arise not just through discourse but in processes of living in which we inevitably engage with the world around us (Casey 2001). Following on from this, as Divya Tolia-Kelly (2006, 213) has argued, theories of affect must recognise the "political fact of different bodies having different affective capacities" in order to avoid ethnocentric universalism. Ideas about affect, emotion and embodiment, cognisant of this critique, have also gained prominence within the field of heritage studies (Madgin and Lesh 2021; Marchant 2019; Smith 2021; Tolia-Kelly et al. 2017; Waterton 2014; Waterton and Watson 2013) to attempt to explore and explain these processes of sense-based cognition within a heritage context. Consequently, heritage (and tourism) is now understood more as an active process or as an embodied performance that comes into being in engagement with people and places, rather than as a static pre-existing thing (for example, Byrne 2014; Harrison 2013; Smith 2006; Waterton 2014; Waterton and Watson 2013).

Although there is considerable research focusing on visitor responses to interpretation in a museum context, including studies of emotion and affect (Micieli-Voutsinas 2017; Schorch et al. 2017; Smith 2014; Waterton 2023; Witcomb 2013), there is limited research on visitor responses to heritage interpretation within the landscape and little that explores affect, emotion and embodiment based on direct engagement with visitors (but see Smith 2014; Smith 2021; Waterton and Watson 2015; Whittington and Waterton 2021). There has also been little direct research on the important question of how the *interpretation* of heritage places mediates and influences our embodied, emotional and affective responses to them (see Clarke and Waterton 2015; Dittmer and Waterton 2019). My research thus not only seeks to explore the discursive aspects of heritage interpretation but to address a gap within the scholarship in regard to the potential impact of different modes of interpretation on visitors' affective responses to it.

This line of inquiry is congruent with research that suggests that emotional engagement with interactive interpretation is important in creating a positive experience that is remembered and is more likely than static interpretation to lead to the acquisition of knowledge, the standard goal of most heritage interpretation (Hughes and Morrison-Saunders 2005; Uzzell 1989). However, rather than seeking to assess the effectiveness of interpretation in terms of empirically measured knowledge acquisition, I explore its effectiveness in emotionally and cognitively engaging the visitor, leading to increased empathy and appreciation for the Aboriginal cultural heritage in place. This approach is supported by work of Emma Waterton et al. (2016) at Kakadu and Uluṟu, which revealed that these places prompted "deep personal reflection" among visitors, leading to "strong feelings of respect and renewed knowledge regarding the importance of Aboriginal and Torres Strait Islander peoples' ties to the landscape and the meaning those ties have for cultural connection and continuation" (Waterton et al. 2016, 9).

Interpretive panels authorised, and in most cases commissioned, by protected area management are a primary *in situ* type of interpretation in the research locales chosen, which comprise Uluṟu-Kata Tjuṯa National Park in Australia's NT and locations within the New South Wales (NSW) Greater Blue Mountains World Heritage Area. While panels are considered accessible to (English speaking) visitors, they have the disadvantage that messages are fixed, necessarily brief and are largely non-interactive. This interpretive media is compared with visitor-controlled media such as audio tours on mobile phone applications, and more interactive approaches such as guided tours conducted by a non-Aboriginal ranger and by Traditional Custodians to determine the degree of visitor engagement. Comparing the impact of different interpretive strategies has allowed me to identify those approaches that invite visitors to engage more fully with a place and its significance to Traditional Custodians, including coming to appreciate the Aboriginal ontology of Country. Relatedly, I investigate whether forms of heritage interpretation developed in consultation with Traditional Custodians and/or developed and delivered by them, lead to higher levels of visitor engagement with, empathy for and self-identified understanding of Aboriginal peoples and their cultures. In so doing, I engage with scholarship that argues that tourism in general, and Indigenous cultural tourism in particular, can function as a catalyst for personal change in the visitor, with broader social and political ramifications (Galliford 2012; Higgins-Desbiolles 2003; Lean 2016; Lloyd et al. 2015).

The question of whether Traditional Custodian control over interpretive media facilitates such "transformations" is an important related issue, including whether it results in a broaching of settler-colonial silences about the impact of the colonisation of Australia on Aboriginal and Torres Strait Islander peoples (see Chapter 7). This has implications for the question of whether Aboriginal cultural tourism can play a role in processes of "Makarrata" or truth-speaking in the Australian context, and so contribute to reconciliation between Aboriginal and non-Aboriginal Australians, as argued by some scholars (Galliford 2010, 2012; Higgins-Desbiolles 2003). Makarrata is a Yolngu word that describes "a process of conflict resolution, peacemaking and justice" (Pearson 2017, n.p.), and literally means "coming together after a struggle" (Appleby and Davis 2018, 504). The word began to be used by the National Aboriginal Conference in the 1970s to refer to a Treaty of Commitment they recommended be entered into between the Australian Government and Aboriginal Nations. More recently, the Uluṟu Convention, in the 2017 Uluṟu Statement from the Heart, has called for a Makarrata Commission to "supervise a process of agreement-making between governments and First Nations and truth-telling about our history" (Pearson 2017, n.p.). As Gabrielle Appleby and Megan Davis (2018, 505) elucidate, "[a] truth-telling exercise of this nature is not directed at determining a single truth necessarily, but at developing public understanding and deliberation about the different experiences of the society's history". I argue that the interpretation of Aboriginal cultural heritage within protected area settings should ideally include a focus on truth-telling, including in relation to sometimes traumatic histories. Without this component, such interpretation is in danger of recreating romanticised, fictitious versions of the past and the present which do not advance the important public understandings that Appleby and Davis advocate.

However, neither are processes of reconciliation well served by settler-Australians engaging in Aboriginal cultural tourism as a way of, consciously or unconsciously, appropriating Aboriginal identities in a settler search for belonging. I explore the meanings visitors attribute to an Aboriginal ontology of Country, including the ways some attempt to integrate it into their own self-concept and apply it to their own relationship with the Australian landscape. I take a critical stance on the issue of visitor "Indigenisation", when engagement with Indigenous cultures moves from respectful appreciation to imitation and potentially, appropriation, of Indigenous identity (Chandler and Reid 2019; Radcliffe 2017; Todd 2016). This position is in line with Wolfe's (2006, 389) argument that while "settler society required the practical elimination of the natives (sic) in order to establish itself on their territory", it "subsequently sought to recuperate indigeneity (sic) in order to express its difference—and, accordingly, its independence- from the mother country", through settler nationalism. Bringing together literature in the fields of critical Indigenous studies, heritage studies, anthropology, geography and tourism, and drawing on research findings arising from consultations with visitors and Traditional Custodians, I hope to contribute to scholarship on the important question of how settler Australians can respectfully engage with Aboriginal ways of being through cultural tourism, while acknowledging that Aboriginal ontologies and epistemologies such as Country are an incommensurable part of Aboriginal and Torres Strait Islander sovereignty.

Identifying the research locales

Three national parks were initially identified as locales to explore the question of how visitors respond to Aboriginal cultural heritage, with several case study areas provisionally selected within each park. These parks were Kakadu National Park and Uluṟu-Kata Tjuṯa National Park in the NT and the NSW Blue Mountains National Park. Following the consideration of my applications to conduct research in the NT national parks by Parks Australia and consultations with Traditional Custodians of all proposed study areas, a final set of research locales was arrived at as follows:

- Uluṟu-Kata Tjuṯa National Park (NT)
  - The Kuniya walk, Mala walk and Lungkata walk close to the Uluṟu monolith
  - Base of the (former) Uluṟu climb, and
  - Uluṟu Cultural Centre.
- The Blue Mountains National Park (NSW)
  - Red Hands Cave, Blue Mountains National Park, Glenbrook
- The Jenolan Caves Reserve, near Oberon and
- Yellomundee Regional Park, Hawkesbury Local Government area.

The Traditional Custodians of Uluṟu-Kata Tjuṯa National Park are the A̲nangu ("the people" in Pitjantjatjara). The Traditional Custodians of Yellomundee Regional Park are the Dharug people, who also have a relationship with Red Hands Cave, Glenbrook, as do the Gundungurra people. The Gundungurra are recognised as the Traditional Custodians of the Jenolan Caves and the Wiradjuri people also have a relationship with the Jenolan Caves.

The parks and case study areas were initially selected because they provide examples of what I regarded as better and poorer *in situ* interpretive practice (Clarke and Waterton, 2015), although some of the case study areas were modified following consultation with Traditional Custodians. My initial assessment of the quality of *in situ* interpretive signage was based on factors such as whether the first-hand perspectives of Traditional Custodians were included in interpretive messages or whether these were omitted. For example, most interpretation in Uluṟu-Kata Tjuṯa National Park dates from 2010 and features first-person accounts attributed to Aṉangu. By contrast, the interpretation associated with Red Hands Cave in the Blue Mountains National Park was installed in 1988 and excludes Traditional Custodian perspectives. Another consideration in locale selection was the hypothesis that varying patterns of white colonisation, types of ownership and management regimes and the different meanings attached to landscapes such as Uluṟu-Kata Tjuṯa National Park and the Blue Mountains National Park in the settler-colonial Australian context might influence the way they are experienced by visitors.

Uluṟu-Kata-Tjuṯa National Park, established in 1958 as the Ayers Rock-Mt Olga National Park, is a relatively small park of 1,326 square kilometres in area, located in Central Australia within the NT, close to the border with South Australia. Following the return of ownership of the park to Aṉangu Traditional Custodians by the Australian Government in 1985, an event colloquially known as "hand-back", it has been jointly managed by Aṉangu and the Australian Government through Parks Australia. The Blue Mountains National Park was declared in 1959 and is now over 2,470 square kilometres in area, located in NSW "between the coastal lowlands of the Cumberland Plain and the Great Dividing Range" (NSW NPWS 2001, 3). The park's easternmost boundary is situated approximately 50 kilometres from the centre of Sydney and in close proximity to its western suburbs. Both Uluṟu-Kata Tjuta National Park and the Blue Mountains National Park are recognised under the United Nations Educational, Scientific and Cultural Organisation's (UNESCO) World Heritage Convention by inclusion on the World Heritage list.

Unlike Uluṟu-Kata-Tjuṯa National Park, the Blue Mountains National Park is managed solely by the NSW National Parks and Wildlife Service (NPWS) rather than being jointly managed with its Traditional Custodians, the Dharug and Gundungurra peoples. However, the Gundungurra have an Indigenous Land Use Agreement (ILUA) with the NSW Government over parts of the Greater Blue Mountains World Heritage Area dating from 20 June 2014. This agreement includes the research locales of Red Hands Cave and the Jenolan Caves. While the ILUA requires the NSW Government to consult with the Gundungurra parties to the ILUA in regard to decisions that affect their cultural heritage within the geographic areas covered by it, this arrangement stops well short of joint management. For example, strategic decisions within jointly managed national parks such as Uluṟu-Kata Tjuṯa are made by a Board of Management with a majority of Aṉangu members, and there are a number of sub-committees of this Board that Aṉangu also participate in with a decision-making role. Additionally, Aṉangu hold the title to Uluṟu-Kata-Tjuṯa National Park following "hand-back", while all NSW national parks continue to be owned by "the Crown". Crown Land is an anachronistic colonial term that refers to "land that is owned and managed by the state government on behalf of the people of NSW" (NSW Government n.d.a).

Uluṟu-Kata Tjuṯa National Park also represents what is understood in Australian geographical classification systems as a very remote location, arguably associated within the white Australian psyche with constructions of "authentic" Aboriginality (Fforde et al. 2013; Muir 2014; Prout and Howitt 2009). By contrast, the Blue Mountains National Park is located on the urban fringe of Sydney, Australia's largest city, and although most Aboriginal Australians now live in urban and regional areas, such settings are paradoxically less likely to be associated with Aboriginal cultures by non-Aboriginal people (Prout 2011). A further point of contrast between Uluṟu-Kata Tjuṯa National Park and the Greater Blue Mountains World Heritage Area is that the former has a later history of white incursion and has not been heavily occupied by non-Aboriginal people.

As alluded to above, changes were made to the choice of case study sites and other aspects of the research plan after consulting with Traditional Custodians and other stakeholders as part of the research's decolonising methodology. These changes were made to ensure that Traditional Custodians had more control over the research process out of respect for their ongoing sovereignty and to enable the research outcomes to have greater relevance to them and to other stakeholders such as Parks Australia (Smith 2021). Although the inclusion of the Jenolan Caves had not initially been planned, it allowed for the study of visitor responses to a locale that has been transformed by tourist visitation since the late nineteenth century into a Europeanised environment, in stark contrast to areas of widely acknowledged Aboriginal cultural heritage such as Uluṟu-Kata Tjuṯa National Park. Although not part of the initial research design, Yellomundee Regional Park, which borders the Blue Mountains Local Government Area, was included at the invitation of a Dharug Traditional Custodian. Incorporating this location into the research plan enabled me to research visitor responses to a longer, more immersive cultural experience in the form of the week-long Culture Camps that are held there, in contrast with other Traditional Custodian led experiences, which varied from several hours to one day in length.

While both the Greater Blue Mountains World Heritage Area and Uluṟu-Kata Tjuṯa National Park are recognised under UNESCO's World Heritage Convention, the latter is acknowledged for both its natural and cultural values,[3] due to the significance of "rock art" galleries and general associative values to A<u>n</u>angu.[4] The Greater Blue Mountains World Heritage Area is currently formally recognised for its natural values alone, despite the significance of this landscape to a number of Aboriginal peoples, including the Dharug and Gundungurra Nations. However, it is important to note that staff of the NSW NPWS interviewed as part of my research acknowledged the important cultural values of the Greater Blue Mountains World Heritage Area to Aboriginal peoples and indicated that they were working to amend the original World Heritage Listing, which occurred in 2000, to include these values. The different classification systems for these two World Heritage Areas in place at the time my research was conducted, and at the time of writing this book, arguably reflect a dualism in UNESCO's World Heritage classification system, which posits nature as an entity that exists prior to and outside of the human realm of culture. A nature/culture binary is also reflected in the way that visitors perceive and experience these parks, with the Blue Mountains National Park primarily valued for its natural heritage (Waterton et al. 2016).

Both the Blue Mountains National Park and Uluṟu-Kata Tjuṯa National Park have high levels of tourist visitation. The Blue Mountains National Park is the most visited park in NSW, recording over 6.4 million visits in 2022 (NSW Government n.d.c), while Uluṟu-Kata Tjuta National Park had 222,118 visitors in 2022 (Australian Government, 2023), down from the pre-pandemic figure of 406,840 in 2019 (Parks Australia, n.d. a). The Blue Mountains National Park receives a high proportion of short-stay visitors engaged in daytrips, while trips of this type are not viable to Uluṟu-Kata Tjuṯa National Park because of the distances and time involved in reaching it due to its remote location. In 2019, the Blue Mountains Local Government Area received 836,000 domestic overnight visitors and 2,624,000 domestic day trip visitors compared with a smaller total of 109,000 international visitors. While international visitor data is not disaggregated by day trip or overnight visit, international visitors stayed on average four nights and domestic visitors two nights (Australian Government, n.d. a). The MacDonnell region, NT, which includes Uluṟu-Kata Tjuta National Park, and the town of Yulara, where visitors to the park stay, recorded 328,000 domestic visitors and 163,000 international visitors in 2019, both of whom stayed on average for three nights (Australian Government, n. d. b).

Unpacking research framing and terminology

There are a number of assumptions underlying the research questions explored and the terminology used in this book that require elaboration. One of these is the meaning of Aboriginal cultural heritage. I understand this term to encompass Aboriginal traditional knowledge in its tangible and intangible dimensions, including

> anything that is written, spoken or created by Aboriginal and Torres Strait Islander Peoples, whether it is a story, a painting, a sculpture, an object, a dance, a song, or music (cultural practices) and any knowledge of their land, culture or kinship that is used to express their cultural identity.
>
> (National Health and Medical Research Council 2018, 18)

It also encompasses specific places that hold meaning to Aboriginal peoples, including contemporary, historic and pre-colonial places, and indeed whole landscapes or Country, due to the beliefs associated with these places, including beliefs about creation stories, creator beings and ancestor spirits. It is a misrepresentation of this concept to view the research locales I have chosen as the focus of my research as places of "Aboriginal cultural heritage" when the entire Australian continent is unceded Aboriginal Country.

The tendency to identify certain demarcated locations as associated with Aboriginal peoples, because, like Red Hands Cave in the lower Blue Mountains, they show clear evidence of prior Aboriginal occupation, has the unfortunate implication of appearing to restrict an Aboriginal relationship with Country to these specific places. This is a fiction of settler-colonialism. The entire Australian landscape is of significance to Aboriginal peoples because of the creation events that took place there, the presence of creator beings and ancestor spirits, and the important relationships that Aboriginal peoples have with the plants and animals within these landscapes (Bird Rose 1996b; Grieves 2008; Moreton-Robinson 2022). Attachments can also exist to other locations that are associated with the experiences of Aboriginal peoples following white colonisation, including places in towns and cities (Harrison 2004; Porter 2010). However, the identification of certain places such as Uluṟu-Kata Tjuṯa National Park with Aboriginal cultural heritage motivates both domestic and international tourist visitation (Abascal et al. 2016; Carr et al. 2016; Ruhanen et al. 2015) and is entangled with visitor beliefs about Aboriginality. My focus on research locales such as Red Hands Cave in NSW and Uluṟu-Kata Tjuṯa National Park in the NT thus reflects the ways that these landscapes are framed within discourses of settler-colonialism, including heritage and tourism discourses.

Australian heritage regimes, through local, state and national heritage classification systems, effectively designate certain places as heritage, including places of importance to Aboriginal peoples. Designations such as "cultural landscape" are applied to locales such as Uluṟu-Kata Tjuṯa National Park within the fields of heritage and geography to acknowledge the human-landscape interaction as reflecting "continuous, dynamic systems of social and cultural life" (Foxlee 2007, 49). However, the term "cultural landscape" is unlikely to be recognised by Aboriginal or non-Aboriginal peoples who are members of the general public, and thus forms part of an expert discourse for describing human-landscape interactions. When applied to places such as Uluṟu-Kata Tjuṯa National Park, through its World Heritage designation as an "associative cultural landscape", such terms have the effect of "repositioning indigenous [sic] peoples outside of their systems of authority and positioning heritage experts as mediators of their heritage" (Baird 2013, 330) (see also Smith et al. 2019). Through processes of naming, accompanied by heritagisation, such places become incorporated into Eurocentric discourses of heritage and tourism. As locales with multiple heritage designations and large tourist numbers, I have chosen Uluṟu-Kata Tjuṯa National Park and the Greater Blue Mountains World Heritage Area as my research foci primarily because these locales give me the opportunity to fully explore these dynamics.

One of the ways that I understand heritage, along with a number of heritage scholars who precede me, is thus as a discourse with material effects: as something that impacts the materiality of places and ways of thinking and speaking about them (Hall 1999; Harrison 2013; Smith 2006; Waterton 2010). As explained further in Chapter 2, my work thus owes a debt to the ideas of Michel Foucault (1981, 2002), and to scholars who have applied discourse analysis to the field of heritage studies (Waterton et al. 2006; Wu and Hou 2015). Heritage, it has been argued, is a discourse that deploys certain romanticised notions of the past to serve the purposes of the present, in ways that reinforce unequal power relations (Lowenthal 1998). The association of heritage with conservative ideas of race and belonging in the British context has been explored by scholars such as Stuart Hall (1999) and has also received attention in more recent scholarly work (Ashley and Stone 2023; Littler 2005).

Heritage, I argue, does similar things in a settler-colonial Australian context, in that it can marginalise Aboriginal and Torres Strait Islander ways of being and knowing, and negatively impact their relationships with places of importance to them. The idea of heritage, however, does not have a single meaning, and may also be deployed by Aboriginal and Torres Strait Islander peoples in legal, policy and media discourses to argue for the protection of their cultural heritage (Byrne 2003a). While some Aboriginal peoples may utilise terms like "cultural heritage" in order to protect places of significance to them in a contemporary Australian context, terms that arise within Aboriginal people's own ontologies and epistemologies, such as the A<u>n</u>angu idea of Tjukurpa, may have a deeper meaning to them. Tjukurpa (pronounced "chook-orr-pa") is "the foundation of A<u>n</u>angu life and society. Tjukurpa refers to the creation period when ancestral beings, jukuritja, created the world as we know it. From this, the religion, law and moral systems were formed" (Parks Australia n.d. b). For this reason, I distinguish between the uncritical deployment of ideas of heritage within Eurocentric and nationalistic discourses, that I argue can be a postcolonising force, and the critical engagement with ideas of heritage within current scholarship (Byrne 2014; Harrison 2012; Waterton 2014; Waterton and Watson 2013).

The term "culture" also figures prominently in this book, both as part of the dyad "cultural heritage" and in reference to the lifeways and extant material and intangible cultures of Aboriginal peoples. I understand culture as a "living process, historical, dynamic and evolving" (UN 2009, 3), expressed in the present lifeways of communities, but also arising from their shared histories and "the persistence of tradition" (Duelke et al. 2005, 201). As Gaynor MacDonald (2001, 182) succinctly explains, culture is those "social, intellectual and materially-orientated practices through which people express to them what it means to be in the world", practices that are never "static or unchanging". Culture, then, is self-defined, part of a group's "social identity" (Cowlishaw 2009, 188) and is both dynamic and historical. However, joining the two terms "culture" and "heritage" together immediately places emphasis on pre-existing material culture because of the popular and historic associations of heritage with property that is handed down (Davison 1991) and with monumental architecture and sculpture (Smith 2006). When the term "Aboriginal" is added to this dyad, the notion of tradition and the deep past is almost inevitably evoked, diminishing the dynamic nature of culture and its material and intangible manifestations (Duelke et al. 2005; Macdonald 2001). As I explore, when visitors travel to certain places seeking "Aboriginal cultural heritage", the tourist journey may consequently become a type of authenticity quest, resulting in shallow engagement and reinforcing various stereotypical beliefs about Aboriginal peoples (see Chapter 5) (Dodson 2003; Fredericks 2013; Gorringe et al. 2011; Grieves 2008; Lambert-Pennington 2012; Maddison 2013; Moreton-Robinson 2016; Porter 2018).

The research locales I have chosen are all landscapes that have been designated by Australian governments as protected areas primarily for the purposes of nature conservation, recreation and tourism, with the exception that Uluru-Kata Tjuta National Park is also formally recognised for its cultural values to Anangu. I focus on protected areas because I am interested in engaging with the range of meanings and associations that these places hold for visitors and the ways that they perceive these places when they experience them first-hand. This includes whether they perceive these places solely through the Eurocentric lens of landscape or, to the extent that this is possible, through the Aboriginal ontology of Country. The concept of Country is complex and "multidimensional" (Wright et al. 2012, 58), and the meanings of this concept to Aboriginal peoples and the ways that it is adapted for use by non-Aboriginal people participating in Indigenous cultural tourism are themes I explore throughout this book. Country is a term used in Aboriginal English "to describe land as a 'nourishing terrain' (Bird Rose 1996b): sentient, sapient ... and intertwined with Indigenous kinship, ancestry, law, language, and culture" (Wright et al. 2012, 58). Country is agentic, and can both sustain life and inflict harm, if the law that Aboriginal peoples believe applies in certain places is not adhered to, as expressed by the Anangu concept of Tjukurpa. Despite the negative impact of colonisation on the relationships that Aboriginal peoples have with Country, for example, as a result of the child removal policies of Australian governments (Haebich 2011), many contemporary Aboriginal peoples, including those living in urban areas, continue to emphasise the importance that Country holds to them (Fredericks 2013; Grieves 2008).

Although Country refers specifically to an Aboriginal ontology or way of being associated with connections to land, ancestral beings and human and non-human kin, the idea that non-Aboriginal people living in or visiting Australia may also form a relationship with Country is promoted by some contemporary Aboriginal and non-Aboriginal scholars (Bawaka Country et al. 2017; Darug Ngurra et al. 2019; Graham 1999). In the later stages of my research, I came to understand the Aboriginal ontology of Country as encompassed by the notion of Indigenous sovereignty, which can be understood as incommensurable, because it is "inherent" and "inextricably connected to being in and of ...[Indigenous] lands" (Moreton-Robinson 2022, 259). This understanding has significant implications for the way that non-Aboriginal visitors may come into relation with Country, as explored in greater detail in Chapters 5–7. Specifically, it places limitations on the extent to which non-Aboriginal people can access this ontology based in part on their differing historically and socio-politically informed positionalities.

Understanding researcher positioning

The importance of identifying the positionality from which one approaches research derives from a recognition that no research is value neutral (Brown and Strega 2005; Saukko 2011). Rather, certain subjective characteristics of the researcher, and the meaning attributed to them within cultural and social contexts, in addition to the researcher's worldview, have a bearing on the research they are conducting (Rose 1997). Factors such as one's gender identity, cultural background, social class and land relations, may, for example, influence how a researcher poses research questions, develops their methodology and interprets research findings. I use the term land relations to refer to the relationship of a person, in this case, a researcher, to the land they live on in a settler-colonial Australian context. For example, a person may be a visitor, settler, or an Aboriginal and/or Torres Strait Islander person, who may be a Traditional Custodian of that land, that is, a person with ancestral ties to that land, or a member of another Aboriginal Nation. A statement of researcher positionality involves being explicit about the multiple factors that constitute one's identity and critically reflecting on how they may influence the research process. Identifying one's positionality is an important aspect of decolonising research, particularly when a settler researcher such as myself engages with issues of relevance to Aboriginal peoples (Nicholls 2009; Russell-Mundine 2012; Stinson et al. 2021).

Anishinabe scholar Kathy Absolon and Saskatchewan scholar Cam Willet (2005, 107) argue that in order to avoid unethical, Eurocentric research, any research undertaken with Indigenous peoples should not make claims to objectivity or neutrality, but requires the researcher/writer to "reveal his or her (sic) epistemological location at the outset through a brief introductory autobiography". They explain that, in order to be ethical, such an autobiography should reveal why one has a stake in the research, since in their view, all research should have personal meaning and significance to the researcher. Importantly, any attempt to undertake decolonising research entails "beginning from our own locations" (Boudreau Morris 2017, 464), and this includes paying particular attention to the land one lives on. Not only this, for white subjects such as myself, acknowledging our racial privilege is necessary to avoid perpetuating exploitative practices through our research. I use the term "white" not to refer to a racial essence but to "...a set of locations that are historically, socially, politically, and culturally produced" (Frankenberg 1993, 6), that accord subjects who conform to this category privilege in settler-colonial Australia.

However, as Catherine Bennett et al. (2022, 9) argue, a "one-time written declaration of our identity" in scholarly work, such as I provide below, is insufficient, and can simply represent a type of performativity, through which the white researcher seeks to demonstrate their own moral worthiness. Declarations of whiteness, as Sara Ahmed (2007, 149) recognises, citing Fine, Weis, Powell and Wong (1997, xi), risk re-centring whiteness and reifying it "as a fixed category of experience", a type of ahistorical essence. But neither is it possible to ignore one's whiteness and the status and privilege it accords. In order to progress declaratory statements such as I make below beyond mere performativity, I have followed through with an ongoing practice of critical engagement throughout the research by utilising auto-ethnography both as an affective method and as a way of integrating critical reflexivity into the research journey.

Reflexivity may be understood as an ongoing practice of critical reflection about oneself in relation to the research, that develops self-awareness about "one's biases, theoretical predispositions [and] preferences" and the impact these may have on the research (Schwandt, 2015, 268). Reflexive practice is "multilayered and sustained" (Rix et al. 2014, 3), encompassing the dimensions of reflection on self, an interpersonal dimension involving reflecting on one's interactions with research participants, and a systems dimension, that examines the ways that institutional policies and practices impact on research processes and outcomes to highlight relations of power. Reflexivity is thus an important aspect of researcher positioning, a mode of analysis and a tool for tracking the changes in a researcher's worldview over time that involves "coming to know the self within the processes of research itself" (Lincoln et al. 2018, 246).

Statement of researcher positionality

I am a white "settler" woman of working-class origins living on unceded Dharug and Gundungurra Country in the Blue Mountains west of Sydney. As someone who identifies as a white female, I recognise that this racial position and gender identity carry with them their own privilege in an Australian context and that "[w]hite women's privilege ... has become normalised because it is grounded in the assumption that the womanness of all women was the same version of womanness experienced by white middle-class women" (Moreton-Robinson 1999, 33). While I am of working-class origins, the version of "womanness" or womanhood I embody is a privileged one based on a white norm.

Although the term "settler" is often used to describe non-Indigenous people who reside on Indigenous lands, and I employ it in this book to describe my own land relations, it euphemises processes of land-theft and genocide on which my ancestors' relationships with land were based. Neither am I entirely comfortable applying the moniker of "invader" to myself, as it implies an active incursion onto another's land. As at least a fourth-generation Australian, this active incursion occurred when my ancestors arrived in the mid-nineteenth century; however, I acknowledge that I am the reluctant beneficiary of it and perpetuate this injustice by making my home on unceded Dharug and Gundungurra land. The above discussion thus serves to highlight the problematic nature of my land relations as a "settler" Australian, in part conveyed by the lack of what I consider adequate or appropriate terms to describe these.

My research focus on the way visitors engage with landscape, Country and Aboriginal heritage is thus to some extent about my own identity and belonging vis-a-vis Australia's status as a once colonial and still postcolonising nation. While I seek to engage with the Aboriginal ontology of Country, I am at the same time aware of the problematic nature of this task for a white "settler" such as myself. Difficulties include the degree to which this ontology can be understood by a non-Aboriginal person and the appropriateness of its use by a "settler", including whether such use represents an ongoing colonisation of Aboriginal knowledges and ways of being. On the other hand, my interest in the ontology of the Country could also be linked with the revitalisation of Indigenous cultures in the late twentieth and early twenty-first centuries (Carlson and Berglund 2021; Kolig 2005), which has to some extent resulted in the mainstream popularisation of such concepts.

In reflecting on one's motivations for conducting research, there is always a degree to which awareness of these is unavailable to us, as we are not transparent to ourselves (Rose 1997). As a "settler", I have previously described my thoughts and feelings about my land relations as characterised by states of undecidedness and ambiguity (Whittington 2020), and the choice of the Blue Mountains as a research location allows me to further explore this positioning. As advocated by Boudreau Morris (2017), I thus literally begin from my own location, considering what it means to live in a place that is also a research location. Although at times uncomfortable, emotional states such as unsettlement have been identified as necessary for "settler" subjects working to decolonise solidarity in partnership with Indigenous peoples (Boudreau Morris 2017; Kluttz et al. 2020; McGloin 2015), and are equally applicable to decolonising research. Critical reflections arising from experiences of unsettlement during the research were recorded in a research journal as a form of auto-ethnography and are integrated into the body of the book at various intervals, appearing most frequently within text boxes.

Limitations of research

The extent to which it has been possible to engage and consult with Traditional Custodians of the study areas constitutes one of this research project's limitations. My consultations provide an insight into the views of the Traditional Custodians who participated in the research, rather than representing a comprehensive investigation of Traditional Custodian perspectives on tourist visitation. Within the parameters of the research, it has also not been possible to fully discuss all Traditional Custodian responses, just as it has not been possible to discuss all responses provided by other participants. The responses of Traditional Custodians are progressively integrated into the book chapters where they align with the chapter's overall theme, providing insight into their distinct perspectives on the research questions. The intentions that Traditional Custodians have for the cultural tours they provide are discussed in detail in Chapter 4. A publication arising from the doctoral research deals in greater depth with the attitudes of the Traditional Custodians of Uluṟu-Kata Tjuṯa National Park, the A<u>n</u>angu, to tourism within the park (see Whittington and Waterton 2021), and future publications may also present Traditional Custodian perspectives.

Additionally, I became aware while conducting the Blue Mountains component of the research that including Aboriginal peoples who lived in the Blue Mountains but were not Traditional Custodians may have been advisable. The impact of the colonisation of NSW on Aboriginal peoples included forced removal and dislocation from traditional homelands, and this means that many Aboriginal peoples currently living in this area do not have ancestral ties to it. This points to a conceptual limitation of my research, as my focus on Traditional Custodians and their relationship with Country does not necessarily encompass the way that Aboriginal peoples who are not Traditional Custodians relate to the land on which they live. This issue was to some extent addressed through the participation of three young Aboriginal women who are not Traditional Custodians in an interview when they visited the research locale of Red Hands Cave (see Chapter 6).

I also recognise that the methodology adopted does not comprehensively cover all forms of heritage interpretation within protected areas. Privately operated guided tours run by non-Aboriginal people were initially excluded from the research to limit its scope to a manageable level. However, as many visitors obtain interpretive information from guides employed by private tour operators – although this interpretation varies widely in quality (Weiler and Black 2003) – at the request of A<u>n</u>angu, private bus tour providers were included in the research within Uluṟu-Kata Tjuṯa National Park as interviewees. I also participated as an observer in a bus tour operated by a large bus tour provider within Uluṟu-Kata Tjuṯa National Park. The necessarily limited scope of research activities that it was feasible to undertake meant that it was not possible to fully explore the impact of privately-operated bus tours on visitors or the type of messages communicated to visitors. This is a gap that could be addressed in follow-up research. Additionally, visitor use of mobile phones (as opposed to mobile phone applications) to download interpretive material from the internet was excluded from the study.

Notes

1 The International Union for the Conservation of Nature (IUCN) defines a protected area as "[a] clearly defined geographical space, recognised, dedicated and managed, through legal or other effective means, to achieve the long-term conservation of nature with associated ecosystem services and cultural values", https://www.iucn.org/resources/conservation-tool/iucn-green-list-protected-and-conserved-areas, accessed 19/07/2022.

2 I use the terms Aboriginal peoples or Aboriginal and Torres Strait Islander peoples when referring to Australian Aboriginal peoples in a general way. The term "Traditional Custodians" is also used to refer to the relationships of Aboriginal peoples with their traditional Country. In recognition that Aboriginal peoples often prefer to describe themselves using terms linked to their language groups and traditional Country, such as Aṉangu, Dharug and Gundungurra, these terms are used when referring to specific groups of Aboriginal peoples. In recognition that the term Indigenous may not be preferred by Aboriginal peoples because it is considered generic, this term is primarily used when discussing Indigenous peoples globally, for example, in the context of Indigenous cultural tourism, https://aiatsis.gov.au/explore/indigenous-australians-aboriginal-and-torres-strait-islander-people, accessed 28/11/2022.

3 Cultural and natural heritage are defined in Articles 1 and 2 of the World Heritage Convention. Cultural heritage is defined as monuments, including architectural works, monumental sculpture, and elements or structures of an archaeological nature "which are of Outstanding Universal Value from the point of view of history, art or science". It is also defined as groups of buildings that meet this criteria and sites, whether "works of man (sic) or the combined works of nature and of man (sic), and areas including archaeological sites which are of Outstanding Universal Value from the historical, aesthetic, ethnological or anthropological points of view". Natural heritage is defined as "natural features consisting of physical and biological formations or groups of such formations, which are of Outstanding Universal Value from the aesthetic or scientific point of view" and "geological and physiographical formations and precisely delineated areas which constitute the habitat of threatened species of animals and plants of Outstanding Universal Value from the point of view of science or conservation" and "natural sites or precisely delineated natural areas of Outstanding Universal Value from the point of view of science, conservation or natural beauty" (UNESCO 2019, 19–20).

4 In 1992, the World Heritage Committee recognised cultural landscapes as included under the above definition of cultural heritage sites, including associative cultural landscapes, included "by virtue of the powerful religious, artistic or cultural associations of the natural element rather than material cultural evidence, which may be insignificant or even absent" https://whc.unesco.org/en/culturallandscape/, accessed 14/10/2021.

2 Visitors being, doing and knowing

The relevance of theory in a settler-colonial context

Introduction

In this chapter, I discuss the philosophical frameworks underpinning my approach, and how these have informed the questions being investigated and the methods adopted in conducting the research. These frameworks are decolonial and postcolonial theory, supplemented by the work of scholars in the fields of critical Indigenous and settler-colonial studies. More-than-representational theory and related theories of affect and emotion are also discussed. Additionally, I utilise Foucauldian discourse theory as both a conceptual framework and a method of analysis. I bring these approaches together in my work as I believe they inform each other, since Foucauldian discourse theory underlies colonial and postcolonial discourse theory and some approaches to critical Indigenous studies, as discussed below. These paradigms share the commonality that they regard reality as to some extent socially constructed through the operations of discourse. I counterpose these to more-than-representational theories that understand meaning as created in the act of doing, to explore how visitors' responses to the interpretation of Aboriginal cultural heritage in place are constituted. I thus seek to understand the way that broad philosophical frameworks such as these may inform visitor responses.

Decolonial theory and related approaches are central to the research because settler-colonialism influences multiple dimensions of social, cultural and political life in Australia, including its spatial dimensions, such as the creation of protected areas and touristed landscapes. It is also crucial for understanding how Aboriginal peoples and their cultural heritage are positioned under settler-colonialism, which I understand to be a "specific mode of domination where a community of exogenous settlers … eliminate or displace indigenous (sic) populations and sovereignties, and constitute an autonomous political body" (Veracini 2019, 1). As Lorenzo Veracini (2019, 1) further explains, as "a specific mode of domination … [settler-colonialism] is especially concerned with space", that is, with the claiming and transformation of space. This has occurred, as Tracey Banivanua Mar (2010) argues, through the creation and management of protected areas such as national parks.

DOI: 10.4324/9781032654973-2

Decolonial theory highlights that one of the ways in which settler-colonialism has claimed Indigenous space is through Eurocentric epistemologies and ontologies that usurp the right to define, name and control Indigenous peoples and places. As I will discuss, decolonial theorists place considerable emphasis on critiquing the whiteness and Eurocentrism of dominant knowledge systems inside and outside of the academy (Grosfoguel 2015; Mignolo 2011; Quijano 2007), in both the majority and minority world.[1] Chapter 3 applies a similar critique to the operations of heritage, environmental and natural resource management discourses in protected area interpretation, on the basis that these are dominant knowledge frameworks that marginalise Aboriginal ontologies and epistemologies. Although decolonial thinkers such as Walter Mignolo (2011) criticise the centrality given to the works of Western theorists such as Michel Foucault (1981, 2002) within academic systems of knowledge, describing their dominance as a form of Eurocentrism, I utilise Foucault's theorisation of discourse to reveal how Western knowledge systems establish their hegemony. As this chapter explains, Foucault argued that rules governing the operations of discourse permit certain things to be said at any one time, while prohibiting other forms of discourse. Not only this, certain statements are accorded the status of truth and, because of this, wield considerable power, appearing as discursive formations across a range of social institutions such as the media, government and the tourism industry. At the same time, I am cognisant of the importance of foregrounding Indigenous ontologies and the theoretical writing of black scholars in a book that sets out to contribute to decolonising heritage and tourism scholarship, and emphasise the contributions of scholars such as Bawaka Country et al (2013, 2015, 2017), Moreton-Robinson (1999, 2003, 2015, 2016, 2022), Paradies (2006), Todd (2016), Tebrakunna Country and Lee (2016, 2022) and Tuck and Yang (2012), among others (McKittrick 2021).

In employing Foucault's ideas and those of subsequent scholars who engage with him, such as Stuart Hall (2001) and Siegfried Jäger and Florentine Maier (2016), I align myself with scholars such as postcolonial theorist Ato Quayson (2000), who argues that it is neither "possible or desirable completely to debunk Western inspired theories" (Quayson 2000, 12), some of which have been adapted by majority world thinkers to their specific contexts and purposes. Moreton-Robinson (2015, 126) similarly argues for the relevance of Foucault's conceptual framework on governmentality in "analysing how white possession, as a mode of rationality, functions within disciplinary knowledges and regulatory mechanisms, defining and circumscribing Indigenous sovereignty in particular ways". Heritage scholar Cressida Fforde (2013, 162) and her colleagues also consider Foucault's idea of discourse pertinent to understanding the persistence of negative beliefs about Aboriginality, which she typifies as a "deficit discourse" in the contemporary Australian context. Within other heritage scholarship, Laurajane Smith's concept of the authorised heritage discourse, as a dominant way of speaking and writing about heritage that "privileges monumentality and grand scale [and] [i]nnate artefact/site significance tied to time depth, scientific/aesthetic expert judgement, social consensus and nation building" (Smith 2006, 11–12), also owes a debt to Foucault.

I am similarly persuaded of the relevance of Foucault's conception of discourse as a set of "practices that systematically form the objects of which they speak" (Foucault 2002, 49) to the question of how visitors respond to Aboriginal cultural heritage in place. This includes the way that visitors apply the Aboriginal ontology of caring for Country to their own relationship with protected areas, potentially appropriating this ontology through a settler-colonial discourse position that seeks to legitimate their claim to Australia (see Chapter 7). Foucault's (1980) theories of knowledge/power are also helpful in understanding the way that ideas of race, including whiteness and Aboriginality, operate within the contemporary Australian context. However, in recent decades it has been argued that there are limits to the usefulness and applicability of representational approaches such as discourse theory (Thrift 2004; Waterton and Watson 2013), particularly in research that focuses on the way that people respond to place and express affect and emotion. Discourse theory, it is claimed, does not account for the interaction between human and non-human worlds in changing us in unanticipated ways, seeming to posit a pre-existing material world, an inert mass, that is acted upon by the human subject. This chapter briefly outlines these debates as put forward by non-representational and more-than-representational theorists, explaining why I also employ more-than-representational theory and affective methodologies as a way of accessing my own and visitors' lived experience of Aboriginal cultural heritage in place.

Taking up this critique of discourse theory, I note that it is also inadequate for expressing Aboriginal ontologies and epistemologies, including the ontology of Country. The concept of discourse cannot convey the agentic nature of Country, which comprises beings and forces that act independently of human subjects, such as the spirits of ancestors, creator beings, malign and beneficent spirits, plants, animals, land, water and air (Bird Rose 1996b, 2005, 2011; Kearney 2009; Moreton-Robinson 2015). In Indigenous and post-human ways of understanding the world (Todd 2016), humans are not the only or primary actors, and ontologies such as Tjukurpa have timeless aspects that sit outside of Eurocentric temporalities, including the temporalities of Foucauldian discourse theory. Indigenous and post-human ontologies also help to explain how not just discourse, but embodiment and affect, constitute the visitor experience, noting that I understand affect as "situated within matrices of power … sensitive to differently positioned narratives" (Tolia-Kelly et al. 2017, 4) rather than as "outside social meaning" (Hemmings 2005, 564). The relevance of these ideas to my research is further elaborated below.

Foucault's theories of discourse, power and knowledge

Rather than conceiving discourse simply as passages of connected speech or writing, as it is understood in linguistics, for Foucault (2002, 54), discourses "do more than use signs to designate things", but, in his much-quoted adage, "systematically form the objects of which they speak". Stuart Hall sums up Foucault's idea of discourse, as both a "group of statements which provide a language for talking about ... a particular topic at a particular historical moment" (Hall, 1992, 291 cited in Hall, 2001, 72), and our practices, or actions, as these are shaped by discourse. As Foucault (1981, 67) puts it "[w]e must conceive discourse as a violence we do to things, or in any case as a practice which we impose on them". A discourse is thus "something that produces something else (an utterance, a concept, an effect), rather than something that exists in and of itself and which can be analysed in isolation" (Mills 1997, 15).

Although Hall typifies Foucault's concept of discourse as a "system of representation" (Hall 2001, 72), as a post-structuralist, Foucault (1981, 73) held firmly to a critique of "the monarchy of the signifier", and did not accept that signs have a fixed, ahistorical meaning, nor was he seeking to uncover meaning hidden within discourse. As a critic of humanism, Foucault (2002, 60) conceived discourse not as the "majestically unfolding manifestation of a thinking, knowing, speaking subject, but, on the contrary, a totality, in which the dispersion of the subject and his discontinuity with himself may be determined". For him, discourse is rather a space exterior to the subject, which "does not pre-exist itself ... [but rather] exists under the positive conditions of a complex group of relations" (Foucault 2002, 49). In its focus on the conjunction of "a complex group of relations" in the present, Foucault's theories of the operation of discourse share commonalities with more-than-representational theory, rather than being diametrically opposed to it.

These relations, it is important to note, are governed by a number of "rules and practices that produced meaningful statements and regulated discourse in different historical periods" (Hall 2001, 72). As discussed below, these rules and practices are mechanisms that govern the operations of discourse, and, importantly, include "three great systems of exclusion which forge discourse, the forbidden speech, the division of madness and the will to truth", with the latter exerting a "pressure and ... a power of constraint ... on other discourses" (Foucault 1981, 55). It is exclusions of this nature, as explained below, that I argue are at work in official interpretive discourses that operate within both Uluṟu-Kata Tjuṯa and the Blue Mountains National Parks. These discursive practices permit certain things to be said and disallow other things, including accounts of the colonial histories of national parks and of the way settler-colonialism has impacted their Traditional Custodians.

A discourse, according to Foucault (2002, 121), is "a group of statements that belong to a single system of formation; thus I shall be able to speak of clinical discourse, economic discourse, the discourse of natural history, psychiatric discourse". Although discrete, these disciplinary or expert discourses can, and often do, interact with each other, and as he makes plain, extend beyond the disciplines to be found in "the statements made and the opinions expressed in daily life" (Foucault 2002, 198). Although Foucault focused on professional discourses, it is clear from the latter phrase that he believed these had a relation with popular discourses expressed in "daily life", although it has been argued that he did not adequately address the relations between the two (Purvis and Hunt 1993, 480 cited in Smith 2004, 64). Smith (2004, 65) supports Purvis and Hunt's position that discourses become ideological, "when they are taken up within, or connected to, systems of domination" and this can explain the way that discourses are "incorporated into lived experience". This insight is important to my argument because I contend that the operations of discourse are not simply apparent in government-authorised expert statements found in interpretive signage, but can also be found in the speech of visitors interviewed in protected area settings.

The relationship of discourse to the material world has considerable bearing on my research, as I argue that the views about Aboriginal peoples and their cultural heritage expressed by visitors both reflect and reinforce dominant discourses in wider circulation in the Australian context. These discourses, through their utilisation within systems of control enacted by the settler-colonial state, have a direct impact on the lived experience of Aboriginal and Torres Strait Islander peoples, including the violence that they continue to experience at the hands of carceral systems. However, it is important to recognise that discourse as conceived by Foucault expresses the way of thinking or state of knowledge at any one time, and thus is historically specific and contingent. Discourses are not universal, ahistorical or atemporal but are "discontinuous practices" (Foucault 1981, 67), existing in ever-changing ways in particular historical moments and locations, often in relation to each other. What this implies is that while some discourses may be both persistent and pervasive, others are subject to change, as the dominance of one eclipses another, such as the growing unacceptability of the Uluṟu climb among visitors to Uluṟu-Kata Tjuṯa National Park (Hueneke and Baker 2009; James 2007; Whittington and Waterton 2021). Nevertheless, Foucault's emphasis on the discontinuous nature of discourse could be regarded as inadequate to explain the ongoing nature of certain systems of oppression, such as those based on racialisation, which decolonial scholars trace to the conquest of the Americas and argue remain dominant (Quijano 2007).

Although conceived by Foucault as contingent and changing, discourses have certain characteristics and rules governing their operation that he argues remain constant and form a basis for analysing them. An important component of this system is what Foucault terms "the statement". He explains that a "sequence of linguistic events" is a statement "only if it is immersed in an enunciative field": in other words, "it is always part of a network of statements, in which it has a role" (Foucault 2002, 111). Mills (1997, 55) understands the statement as those speech acts and texts that "make some form of truth claim" and this is what gives these "sanctioned statements ... institutionalised force", causing them to have "a profound influence on the way that individuals act and think". Not only this, for Foucault, the "corpus of already formulated statements and the capacity to invest this discourse in decisions, institutions, or practices – is in fact confined ... to a particular group of individuals" (Foucault 2002, 75–6), based on discipline and profession.

There is an exclusivity associated with sanctioned statements, such as those, for example, that expert commentators make about Aboriginal peoples in government-authorised interpretive signage in protected areas. It is the expert authority of such statements that gives them the status of truth, even when they exclude Aboriginal peoples from their development, as will be discussed in relation to Red Hands Cave in the Blue Mountains National Park (Chapter 3). In this case "[t]he manifest discourse ... is really no more than the repressive presence of what it does not say; and this 'not-said' is a hollow that undermines from within all that is said" (Foucault 2002, 28). The way that the gaps and silences associated with excluded "others" undermine official protected area interpretive discourse is a theme that I engage with throughout this book.

Another important term developed by Foucault as a way of explaining the operations of discourse is "discursive formation". As Foucault (2002, 41) explains, "... whenever, between objects, types of statement, concepts, or thematic choices, one can define a regularity (an order, correlations, positions and functionings, transformations), we will say ... that we are dealing with a discursive formation". Thus, as Hall (2001, 73) elucidates, a characteristic of discourse is that it can be pervasive, appearing "... across a range of texts, and as forms of conduct, at a number of different institutional sites". When these discursive events share key characteristics, they are understood to belong to the same discursive formation. A discursive formation "can be detected because of the systematicity of the ideas, opinions, concepts, ways of thinking and behaving ... formed in a particular context" and the effects of this systematicity (Mills 1997, 15). Thus, for example, and as is argued herein, discourse fields such as tourism, heritage and the media, share certain positionings in regard to Aboriginal identity, including what constitutes an "authentic" Aboriginal person, and the pervasiveness of this signals the existence of a discursive formation.

As alluded to above, sanctioned statements and discursive formations tend to exclude or limit certain types of discourse. Thus discourse, within certain places and times, regulates what can and cannot be said and this extends to social practices and behaviour. As Mills (1997, 63) explains, Foucault theorises that what we want to say and write tends to fall "... within fairly predictable and restricted sets of parameters" (or discursive limits), and these, according to Jäger and Maier (2016, 121), "constitute the borders of what is not sayable". These discursive limits are typically sanctioned by some type of social institution. Discourse may thus serve to exclude, silence and marginalise ways of being and expression that are outside of its parameters.

Jäger and Maier note that these discursive limits can be identified in text and speech through various signals such as rhetorical devices, including "direct prescriptions, relativizations ... allusions and implicatures" (Jäger and Maier 2016, 121).

Mechanisms through which these exclusions occur include what Foucault terms prohibition, or taboo, by which certain subjects are rendered off limit for discussion in particular societies at different times. He also regarded the division between the type of knowledge that is considered true and that which is considered false to result in the exclusion of the latter from the status of knowledge. The way that Western scientific knowledge is considered true and factual, while Aboriginal accounts of the creation of landscape features such as the Uluru monolith by creator beings is considered myth and therefore false, is an example of this type of knowledge valuing. Foucault (1981, 55) regards the marginalisation of discourses excluded from the status of truth to be one of the key ways in which "the three great systems of exclusion" referred to above "forge discourse".

In his later work, Foucault became increasingly interested in the relations between knowledge, power and the body, particularly the way that knowledge is "applied to the regulation of social conduct in practice" (Hall 2001, 75). This idea of knowledge is comparable to a reified dominant discourse, one that has taken on the status of universal, ahistorical "truth". A discursive formation may thus sustain a "regime of truth", which has real consequences, "and will become 'true' in terms of its real effects" (Hall 2001, 76). Consequently,

> power produces knowledge ... power and knowledge directly imply one another; ... there is no power relation without the correlative constitution of a field of knowledge, nor any knowledge that does not presuppose and constitute at the same time power relations.
>
> (Foucault 1984a, 175)

An example of power/knowledge relevant to my research is the way that certain pseudo-scientific (eugenicist) settler-colonial beliefs about "authentic" Aboriginal identity took on the status of knowledge in certain periods of Australian history and had a direct impact on the lives of Aboriginal Australians through various *Aboriginal Protection Acts* (Armitage 1995; Carlson 2016). Although this legislation no longer exists, beliefs about "authentic" Aboriginal identity continue to circulate in popular, media and tourism discourses, including, as will be discussed, figuring prominently in the discourse of some visitors interviewed as part of this research.

Foucault's ideas have been influential on generations of thinkers, some of whom are also interpreters of his ideas, such as Hall, whom I cite in this chapter. Two contemporary scholars, Siegfried Jäger and Florentine Maier (2016, 111) describe themselves as building on the work of literary and cultural theorist Jürgen Link and his team at the University of Dortmund, who operationalised Foucault's ideas "to analyse discourses and their power effects" as a form of critical discourse analysis. Critical discourse analysis (CDA) has been defined as

> a programme of social analysis that critically analyses discourse - that is to say language in use - as a means of addressing problems of social change. [It]...is founded in the idea that the analysis of discourse opens a window on social problems because social problems are largely constituted in discourse.
>
> (Scollon 2001, 140)

While scholars such as Ruth Wodak (2001) and Norman Fairclough (2003) employ a form of CDA that focuses on analysing the linguistic structure of texts, Jäger and Maier (2016, 119) "aim to identify the knowledges contained in discourses and dispositives, and how these knowledges are connected to power relations in power/knowledge complexes".[2] Not only this, they explain that they extend Link's work by including "non-linguistically performed action, including the creation of material objects" (Jäger and Maier 2016, 111) as part of their analysis.

I employ the approach of these theorists in analysing protected area interpretation by identifying discursive formations or strands in discourse fragments across different texts and discourse planes, and pointing to the material affects of these discourses. A text may, for example, be a government policy statement as well as an interpretive sign, a museum display or an interview transcript. A discourse plane refers to "various social locations from which speaking takes place" (Jäger and Maier 2016, 123). These can also be understood as settings, for example, politics, tourism, the mass media, government administration and so on. Discourse planes are historically and temporally specific, influence each other and relate to each other. They also consist of different sectors, for example, mass tourism, versus niche tourism such as cultural/heritage tourism, which may be interwoven with each other.

Linked to the concept of discourse planes is that of discourse positions, which are the "ideological positions from which subjects, including individuals, groups and institutions, participate in discourse" (Jäger and Maier 2016, 124). Certain discourse positions may saturate particular discourse planes, such as the tendency of a Western scientific discourse of natural resource management to figure prominently in protected area interpretation, marginalising Aboriginal ways of caring for Country (see Chapter 3). Jäger and Maier (2016) explain that hegemonic discourse positions tend to be more homogeneous, but that other discourse positions "are homogeneous only in their core and become diffuse with regard to less central issues" (Jäger and Maier 2016, 125). Entanglement of separate discourse strands or "discursive knots" can be a feature of hegemonic discourses, as this can have the effect of adding weight to certain "regimes of truth". Similarly, collective symbols or cultural stereotypes serve to link up discourse strands by means of catachresis, which is a technique for establishing "connections between statements, bridging contradictions and therefore increasing plausibility" (Jäger and Maier 2016, 123). One such collective symbol arising in visitor interviews is the way in which some visitors labelled protected areas such as Uluṟu-Kata Tjuṯa as "our own backyard", unconsciously enacting the "possessive logics of patriarchal white sovereignty" (Moreton-Robinson 2015, xi) over parts of Australia very distant from their homes on the south-eastern seaboard.

Although discourse analysis is a qualitative approach, Jäger and Maier (2016, 126) discuss the usefulness of integrating quantitative elements, such as the frequency with which particular statements occur. This enables the identification of "focal issues in discourse strands", or "statements that have the character of slogans" and may therefore be associated with prejudices. Jäger and Maier (2016, 124) also discuss the significance of "discursive events", defined as such because they "appear on the discourse planes of politics and the mass media, intensively, extensively and for a prolonged period of time". They explain that discursive events are important "because they influence the future development of discourse". One such event that formed part of my research was the closure of the climb of the Uluṟu monolith in October 2019. The power of this discursive event was apparent in visitor responses to its announcement by Parks Australia, which was evident in May 2019 when I conducted research in the park (Whittington and Waterton 2021).

Although foundational to much subsequent scholarship, Foucault's theories of discourse, power and knowledge, including their adaptions by contemporary scholars, are not without their critics (Carpentier 2018; Jørgensen and Phillips 2011). For example, his concept of power has been criticised as being overly relativistic and based on a male model, failing to take into account the structural operations of power on the basis of sex, race and class (Buker 1990). Additionally, Mills (2003, 123) explains that Foucault "has located resistance within power itself, thus denying the agency of those who oppose oppressive regimes". While Foucault does not deny the existence of the real, Mills explains that, according to Foucault, "what we perceive as significant and how we interpret objects and events … is dependent on discursive structures" (Mills, 1997, 46). Because of his views about the powerful role of discourse in influencing thinking and indeed constituting the social world, his theories could be seen to constrain human agency and freedom, although contemporary interpreters of Foucault such as Jäger and Maier (2016, 112) argue that discourse is always mediated by "intervening active subjects in their social contexts as co-producers and co-agents of discourses". As argued earlier in this chapter, Foucault also had no regard for the agency of the non-human. These critiques suggests that in applying Foucault's ideas, it is important to clarify one's position in relation to claims that discourse is all-encompassing, to be clear about the gendered and racialised specificities of the subjectivities that one is discussing and to highlight the limitations of his theories for explaining Indigenous ontologies such as Country.

Non-representational and more-than-representation theorists, and theorists of affect, discussed below, argue that discursive approaches fail to take adequate account of the dynamism of social and corporeal worlds, including the social world of human interaction. Other contemporary lines of social theory, such as those found in post-humanism, are critical of older theoretical approaches that they consider place undue emphasis on the human subject to the exclusion of the non-human (Braidotti 2019). Furthermore, decolonial scholars have termed Foucault "Eurocentric" (Grosfoguel 2015, 37), viewing him as part of the intellectual tradition of the minority world that has become a dominant orthodoxy globally within universities. In this vein, Mignolo (2011, 279) argues that Foucault's notions of epistemic breaks and paradigmatic changes relate to the history of Europe, and should be understood regionally, not as a universal. This critique suggests that the applicability of Foucault's concepts to non-Western contexts should be tested on a case-by-case basis.

Postcolonial and decolonial theory

While I argue for the usefulness of Foucault's concept of discourse in describing some of the processes through which Aboriginal cultural heritage in protected areas acquires meaning to visitors, postcolonial and decolonial theory is indispensable in explaining the dispositions of power/knowledge in settler-colonial Australia. It is in this context that the idea of heritage itself and the way it operates is postcolonising, if it perpetuates systems of colonial power, such as control of knowledge about Aboriginal peoples and (mis)representation of their cultural heritage.

Emergence of the term "postcolonial" predates that of "decolonial", as it arose in anti-colonial political movements in Asia, Africa and South America from the mid-twentieth century onwards. Pramod Nayar (2010, 1) explains that leaders and intellectuals within these movements generated ideas that "developed into a body of thought in academic practices". This body of thought includes postcolonial theory, arising from the work of diasporic academics from the Middle East and South Asia such as Edward Said (1978), Gayatri Spivak (1988) and Homi Bhabha (2004) (Bhambra 2014). Quayson (2000, 2) provides a useful "working definition" of postcolonialism, as

> a studied engagement with the experience of colonialism and its past and present effects, both at the local level of ex-colonial societies as well as at the level of more general global developments thought to be the after-effects of empire.

It is thus both about conditions under imperialism and colonialism, and those that come after the historical end of colonialism, making it relevant to countries such as Australia both historically and in the contemporary period.

However, Moreton-Robinson (2015, 10), observes that Australia is not postcolonial in the same way as countries such as "India, Malaysia and Algeria can be said to be" because these countries do not have a dominant white settler population. As she explains, "[i]n Australia the colonials did not go home, and 'postcolonial' remains based on whiteness" (Moreton-Robinson 2015, 10). She argues for theorising the present situation in Australia not as postcolonial but as "postcolonising", to signal that colonisation is still in process. Her use of the term "postcolonising" is also intended to signal that Indigenous/white settler colonial societies such as Australia are different from other postcolonial nations, such as India, where movements of independence resulted in the departure of colonial powers.

Without this theoretical intervention, the specific experiences of Indigenous peoples in settler nations such as Australia, Canada and the United States, for example, are elided and made invisible, including that this colonisation is in effect continuing through the ongoing occupation of Indigenous land, and systems of control and management of Indigenous peoples, including tourism and heritage. For example, the heritagisation of places such as Uluṟu-Kata Tjuṯa National Park through its inclusion within national and international heritage systems has contributed to the attraction of this place to both national and international visitors. The presence of hundreds of thousands of visitors in this small park each year benefits A<u>n</u>angu by contributing to the local economy but also disadvantages them by making important places such as the Mutitjulu waterhole at the base of Uluṟu, less accessible to them. Not only this, visitor practices such as climbing Uluṟu, now banned, and photographing A<u>n</u>angu without their permission and also photographing culturally sensitive parts of Uluṟu, despite signage prohibiting this practice, continue to cause significant personal and cultural offence to A<u>n</u>angu (Whittington and Waterton 2021). The theme of colonisation still in progress is also one I explore in this book through a discourse analysis of state-sanctioned interpretation within the research locales (Chapter 3). It is foundational to my analysis of visitors' spoken discourse, including my identification of the stereotypes about Aboriginal identity that influence visitors' travel motivations and encounters explored in Chapter 5. The theme of colonisation still in progress also figures prominently in Chapter 6, where I explore the role of "settler structures of feeling" (Rifkin 2011, 350) in producing visitor responses to both Aboriginal peoples and protected area landscapes.

Influenced by the writings of Foucault, postcolonial discourse analysis emerged as "an analytical-critical approach that treats colonial writing, arts, legal systems, science and other socio-cultural practices as racialized and unequal where the colonial does the representation and the native (sic) is represented" (Nayar 2010, 25). It is thus "a method of interpreting, reading and critiquing the cultural practices of colonialism" based on the idea that "the exercise of colonial power is also the exercise of racially determined powers of representation" (Nayar 2010, 25). In fact, Nayar (2010) argues that the racial discourses of colonialism necessarily facilitated the operations of colonial relations of power, such as the ordering and controlling of colonised populations. Thinkers such as Edward Said (1978) and Mary Louise Pratt (1992) have thus been described as colonial discourse theorists, as they primarily interpreted colonial writings, while Homi Bhabha is best summed up as a postcolonial discourse theorist because he was "more concerned with the effects the colonial enterprise had on current social structures and discursive formations" (Mills 1997, 94).

Edward Said's *Orientalism*, published in 1978, is seen by Williams and Chrisman (1994, 5) as "single-handedly inaugurat[ing] a new area of academic inquiry": that of colonial discourse analysis. This work focuses on the way the West "produced and codified knowledge about non-metropolitan cultures ... under colonial control" (Williams and Chrisman 1994, 5), including what Said termed "the Orient", as both "a figment of the Western imagination for consumption in the West, and also as a means of ... imperial domination" (Quayson 2000, 4). Countries of Asia and the Middle East were collapsed into this term, erasing their cultural particularity and differences. Said demonstrated the way in which a range of texts "encoded formulaic representations, stereotypes and cliches about Oriental religions, cultures and peoples as part of a larger colonial project ... with material-political consequence" (Nayar 2010, 14). Orientalism restricted how the countries thus labelled could be thought of and written about, establishing the West as the binary opposite or "other" of the East as a justification for the colonial project (Quin 2017).

These self-perpetuating essentialist constructions of the Orient include the belief that the Oriental was not only backward but intrinsically "... irrational, depraved (fallen), childlike, 'different'" (Said 1978, 48). In contrast, the European was considered "... rational, virtuous, mature, 'normal'" (Said 1978, 48). Oriental "others" thus required the intervention of the colonial government because they could not effectively govern themselves. Orientalism had not only a narrowly discursive or textual aspect but also a "material and institutional" (Nayar, 2010, 19) dimension, as a system of representation that responded to the demands of the Imperial Age. These ideas are introduced into my discussion as a number are reflected in the paternalistic discourses of visitors about "helping" Aboriginal peoples, and judgements made about them from a position of assumed white superiority, discussed in Chapters 5 and 6. Although the Imperial Age has passed, and Australia is no longer a British colony, my research suggests that a number of colonial discourses that "other" Aboriginal peoples have persisted in "postcolonial" Australia.

However, it is important to understand that Aboriginal and Torres Strait Islander peoples have challenged and continue to dispute and disrupt colonial processes of "othering" that frame them as "the poorer, the lesser, the most in need of correction, discipline and welfare" (Evans et al. 2022, 18). To the contrary, Dharug woman Jennifer Evans, writing with tebrakunna country and trawlwulwuy woman Emma Lee (2022, 18) explains that Indigenous women "do not see ourselves as the Other ... instead we are central to the stories of deep time, world histories and shaping of our societies today". Presenting such processes of "othering" as all-encompassing highlights the inadequacies of Said's (1978) work when applied to the contemporary Australian context, since it does not align with Aboriginal people's lived experience as subjects with agency, and not passive objects at the mercy of colonial processes.

The key postcolonial thinker Homi Bhabha (2012) departed significantly from Said's understanding of the efficacy and impermeability of colonial discourse, which is an area of critique directed at Said by scholars like Quayson. Bhabha argued that colonial discourse was "internally flawed, contradictory ... and often *failed*" (emphasis in original, Nayar 2010, 26), leaving it open to subversion by the colonised subject. For example, although the British imperialist encouraged those subjected to colonisation to emulate or mimic the coloniser, "mimicry, articulates those disturbances of cultural, racial and historical difference that menace the narcissistic demand of colonial authority", creating an "ambivalence [that] disrupts its authority" (Bhabha 2012, 126). Not only this, the coloniser became in a sense dependent on the colonised, as the former based their identity on positioning themselves as different from the racialised colonial "other". As a consequence, Bhabha concluded that colonial identity was not stable or fixed, but relational and that colonial discourse was ambivalent and therefore unstable, because the colonial both feared and desired the colonised "other".

The ambivalence of colonial discourse also allows what appeared to be the subservient mimicry of the British coloniser by the colonised to in fact be subversion and resistance to colonialism. Bhabha (2012) defined this state of being as hybridity, positioning the "native" and migrant in a third space, from which colonial discourse can be subverted. As Bhabha (2012, 72) explains, "the meaning and symbols of culture have no primordial unity or fixity; ... even the same signs can be appropriated, translated, rehistoricized and read anew". Visitors' discourse about the claimed right to climb Uluṟu, for example, as part of a discourse of assumed rights to national and international heritage places and public areas like national parks, can thus be countered by Aṉangu who argue that their ontology of Tjukurpa is primary in that place, and that therefore climbing should be (and has been) prohibited. If colonial discourse were impermeable there would be no space for Aṉangu to make these claims and have them enacted through the closure of the climb.

However, postcolonial thinkers such as Bhabha are not without their critics, since some scholars find fault with the obscurity of their writing and their debt to Western psychoanalytic thinking (Dirlik 1997; Parry 2004). A number of the critiques of postcolonialism have come from scholars dubbed decolonial thinkers, who, while emerging in the same historic period, have a decidedly different disciplinary and theoretical orientation. While postcolonial theory arose from the work of diasporic scholars from South Asia and the Middle East, and referred back to these locations, decoloniality emerged from the work of diasporic scholars from South America and referred back to this location and the imperial powers relevant to it, such as Spain. Unlike postcolonial theorists, who focused on events of the nineteenth and twentieth centuries and their aftermath, decolonial scholars trace the origins of colonialism back to the colonisation of the Americas from the fifteenth century onwards. While most postcolonial theorists were based in literature and cultural studies disciplines, decoloniality had its origins in world systems theory, development and underdevelopment theory and the Frankfurt School critical social theory tradition (Bhambra 2014). It is associated with key proponents such as Gurminder Bhambra (2014), Maria Lugones (2007), Walter Mignolo (2007, 2011), Anibal Quijano (2007) and Ramón Grosfoguel (2015).

Decolonial thinking understands coloniality in fundamentally different ways to postcolonial theory, as "long-standing patterns of power that emerged as a result of colonialism, but that define culture, labor, intersubjective relations, and knowledge production well beyond the strict limits of colonial administrations" (Maldonado-Torres 2007, 243). Coloniality is not viewed as the aftermath or a residual form of colonialism, but rather an ongoing structural social and economic feature of global dynamics based on the creation of ideas of race, and the exploitation of the labour of supposedly inferior races, starting with the conquest of the Americas. In regard to settler-colonial societies such as the United States, Canada, and Australia, decolonial scholars such as Patrick Wolfe argue that settler-colonialism should be understood as "a structure, not an event" (Wolfe 1999, 2), pointing to the way that settler-colonialism is not only ongoing but fundamental to the ordering and functioning of contemporary Australian social, cultural and political systems. This argument is crucial to this book in the sense that I claim that transforming Aboriginal Country into protected landscapes for the purposes of conservation, recreation and tourism represents one way that settler-colonials have laid claim to the entirety of the Australian landscape.

The work of decolonial theorists is also important to my arguments due to the critique they direct against mainstream theory's reliance on Eurocentric epistemologies, primarily postmodernism and poststructuralism, and their call to revalue marginalised Indigenous knowledges. In response, I attempt to foreground the perspectives of not only Indigenous scholars such as Aileen Moreton-Robinson (2015) and Zoe Todd (2016), but the Australian Aboriginal ontologies of Country, relationality and Tjukurpa (Dudgeon and Bray 2019; Grieves 2008; Martin and Mirraboopa 2003; Tynan 2021). The work of Linda Tuhiwai Smith (2021) on decolonising research methodologies and institutions is also central to the methodology used in undertaking the research on which this book is based (see Chapter 8). In foregrounding the work of these scholars, I respond to the critique of the Western academy by decolonial thinkers such as Grosfoguel (2015, 23–24), who characterise the knowledge of "the *Westernized University*" as Cartesian, and as masquerading under a false universalism (emphasis in original).

Rejecting the idea of the Cartesian subject as a mind that functions separately from a body, decolonial theorists emphasise the situatedness of all knowledge in place and time and in particular bodies. They also argue that knowledge is produced through social relations, rather than solipsistically, via the internal monologue of the Cartesian subject. As a feature of its situatedness, decoloniality is also concerned with "global equality and economic justice" (Mignolo 2011, 274), which is why Maeso and Arujo (2015, 19) call for "an approach to the decolonization project as a practice that it is always engaged with profound political and cultural change". Decolonial thinkers are not only concerned with global economic justice. Central to their arguments is the premise that the colonial domination of South America constituted "a new world order, culminating, five hundred years later, in a global power covering the whole planet" (Quijano 2007, 168). While "Eurocentred colonialism" has now been replaced by "Western imperialism", the colonial structure of power, Quijano contends, has an enduring legacy in producing "social discriminations which later were codified as 'racial', 'ethnic', 'anthropological'" (Quijano 2007, 168), which correspond to contemporary patterns of race-based exploitation. He describes the original process of colonisation as one that was simultaneously based on the exploitation of resources, the knowledge of the colonised as to how to access these, the products of their labour, and the systematic repression of the beliefs, ideas and knowledge of the colonised not necessary to this exploitation.

This domination impacted the culture and imaginary of the colonised, a process Quijano (2007, 169) terms "cultural Europeanisation" (Quijano 2007, 169). He also makes the important point that while European coloniality was establishing itself, "the cultural complex known as European modernity/rationality was being constituted" (Quijano 2007, 172). The European, he argues, took on the role of the rational Cartesian subject, and other cultures could only be "'objects' of knowledge or/and of domination practices" (Quijano 2007, 174). Simultaneously, the modernist project conceived history as an "evolutionary continuum from the primitive to the civilized; from the traditional to the modern; from the savage to the rational" (Quijano 2007, 176). Quijano thus identifies "epistemological decolonization" as a vital first step in the "destruction of the coloniality of world power" (Quijano 2007, 177). These ideas are important in interpreting my research findings because the national park-based interpretation of Aboriginal cultural heritage continues to prioritise Eurocentric epistemologies or knowledge systems, and thus, I argue, requires epistemic decolonisation.

Mignolo develops Quijano's critique of modernity/coloniality, taking up his call for "epistemological decolonization". He argues that the "geopolitical locations of knowledge" should be acknowledged, but also seeks to affirm "those modes and practices of knowledge that have been denied by the dominance of particular forms" (Bhambra 2014, 118). Mignolo (2011, 274–5) calls for "delink[ing] from ... theological (Renaissance) and egological (Enlightenment) politics of knowledge" and in its place a turn to "border thinking" and to "the reservoir of ways of life and modes of thinking that have been disqualified by Christian theology since the Renaissance". Border thinking, he explains, means "dwelling and thinking in the borders of local histories confronting global designs, thereby enacting a form of "epistemic disobedience" (Mignolo 2011, 277). Indigenous ontologies and epistemologies, that I argue should be central to the management of national parks and the interpretation of Aboriginal cultural heritage, could be regarded as just such a form of border thinking (Chapter 3).

While Mignolo calls for "border thinking" and Enrique Dussel (2012, 33) seeks to champion "peripheral cultures" that "Western, metropolitan, and Eurocentric culture" has labelled "unworthy, insignificant, unimportant and useless" (Dussel 2012, 42), these epithets of marginality may not be in keeping with the ways in which Indigenous peoples perceive themselves. Labelling non-Western cultures "peripheral" could be seen to reinstate a centre/margin dichotomy that does not reflect Indigenous peoples' understandings of their place in the world. Speaking as Indigenous women, Jennifer Evans and Emma Lee (2022, 22) assert, "we are not a 'marginalized perspective', but a site of empowerment". Consequently, decolonial thinking must not only reject a devaluing of what has been labelled "tradition" within a rhetoric of racialised modernity (Mignolo 2007, 242) but engage in a deep process of "epistemological decolonisation" of the terminology adopted within its own discourse.

The project of far-reaching conceptual decolonisation is taken up by tourism scholars Donna Chambers and Christine Buzinde (2015), who discuss the need to reclaim and utilise Indigenous and non-Western ontologies and epistemologies to truly decolonise the Western academy. As postcolonial scholar Bagele Chilisa (2012, 9) explains, decolonisation must extend to "different but entwined types of colonialism", including "political colonialism" and "scientific colonialism", terms she uses to refer to the imposition of the colonisers' ways of knowing upon those subjected to colonisation. Utilising Indigenous cultural protocols and methodologies such as yarning, as employed in the research upon which this book is based, is an important way to achieve epistemic decolonisation, particularly when conducting research with Indigenous peoples about places of significance to them (Bessarab and Ng'andu 2010; Smith 2021). Further, I apply this critique of modernity and the scientific colonialism of Eurocentric knowledge systems to the modes of interpretation authorised by protected area management authorities and also to visitor narratives that negatively characterise A_n_angu as backward rather than modern.

Despite its call for epistemic decolonisation, decolonial theory is closely linked to an activist agenda, as it sees the goal of scholarship as active decolonisation, rather than simply explicating the way that colonial paradigms function. For Unangax̂ scholar Eve Tuck, writing with William Yang (2012, 1), "decolonization is not a metaphor" but "brings about the repatriation of Indigenous land and life". Treating decolonisation as a metaphor that can be used in a range of contexts gives it a rhetorical rather than activist function, as part of what they term settler "moves to innocence" (Tuck and Yang 2012, 3). In their view, one shared by Moreton-Robinson (2015) amongst others, colonisation, particularly in a settler-colonial context, is about the theft of Indigenous land, which is "remade into property [and a resource] and human relationships to land are restricted to the relationship of the owner to his property" (Tuck and Yang 2012, 5).

Tuck and Yang contend that "decolonization in the settler colonial context must involve the repatriation of land simultaneous to the recognition of how land and relations to land have always already been differently understood and enacted" (Tuck and Yang 2012, 7). Similarly, Hage (2000, 38) explains racist practices as "categories of spatial management" as seen, for example, in the way that white Australians claim the power to define who belongs in the neighbourhood and nation. Arguments about white territorial claims and the repatriation of Indigenous land signal the importance of both land ownership and land relations in the settler-colonial polity. These themes emerge in Chapter 6, which deals with visitors' affective responses to Aboriginal cultural heritage in place, revealing the persistence of the "possessive logic of patriarchal white sovereignty" (Moreton-Robinson 2015, 176) even among visitors who categorise the white invasion of Australia as morally wrong.

More-than-representational theory, affect and emotion

While postcolonial theory engages closely with discourse, and decolonial theory with the power dynamics of postcolonising societies, "more-than-representational" theory seeks to move away from approaches that work to uncover meaning that, it is assumed, is already present in discourse. The term "more-than-representational" was first used by Hayden Lorimer in 2005 to distinguish his approach from that of non-representational thinkers such as Nigel Thrift (1996, 1997, 2004, 2008), John Dewsbury et al. (2002), Ben Anderson and Paul Harrison (2010), David Crouch (2010) and Steve Pile (2010). Non-representation theory emerged in the 1990s within British geography to counter a perceived overemphasis on representation within this field and is closely associated with works by Nigel Thrift such as *Spatial Formations* (1996) and his book chapter "The Still Point" (1997) in Pile and Keith's edited collection, *Geographies of Resistance*. It is a branch of poststructuralism closely linked with the discipline of cultural geography, interested in "how life takes shape and gains expression in shared experiences, everyday routines, fleeting encounters, embodied movements, precognitive triggers, practical skills, affective intensities, enduring urges, unexceptional interactions and sensuous dispositions" (Lorimer 2005, 84).

It is thus typical of what Edward Casey (2001, 684) identifies as the tendency of contemporary philosophical thought, especially phenomenology, contra René Descartes, "to contest the dichotomies that hold the self apart from body and place". Proponents of this approach argue that an over-reliance on theories of representation prevents us from capturing the liveliness that characterises the interaction between the human and non-human in space and place, at points in time. In regard to applying such approaches to research methodologies, proponents seek to move away from methods that emphasise textual analysis, in favour of empirically following and describing how people as embodied beings behave/interact in place/space and subjective accounts of their experiences (Lorimer 2005). However, Lorimer refers to the qualifier "non" in the term "non-representational" as an "unfortunate hindrance" (Lorimer 2005, 84), arguing that the "two realms of research are not irreconcilable" (Lorimer 2005, 86).

Emma Waterton (2014) takes a somewhat different position, contending that non-representational theory, as set out in the work of Thrift, "is not an approach against representation", but one that understands representation as an action that is performed (Dewsbury et al. 2002, 438, cited in Waterton 2014, 826). Meaning-making thus occurs in a performative context, rather than exclusively via representational means such as discourses and semiotics (Anderson and Harrison 2010). The term more-than-representational theory is useful for making clear that this approach is compatible with representational methodologies such as discourse theory, to the extent that discourse is understood as performed and enacted. Margaret Wetherell (2013), a social psychologist working in the area of discourse analysis, also disputes that there is any inherent contradiction in research methodologies that seek to combine discourse analysis of speech with non or more-than-representational approaches such as the study of affect. In fact, she argues that "[a]n obvious entanglement for human social actors occurs routinely between embodied states and the semiotic" (Wetherell 2013, 353), in the sense that we communicate primarily through speech acts.

More-than-representational theory has been taken up by scholars such as Waterton and Watson (2013, 554) as a promising direction in the field of heritage studies to assist in "think[ing] through the ways in which more elusory everyday practices and processes intersect with the cultural world". As they note, "[t]hese sorts of approaches gesture towards the centrality of feelings, emotions and affect, and echo the language of phenomenology by foregrounding concepts such as performance and embodiment" (Waterton and Watson 2013, 554). More-than-representational theory's emphasis on embodied encounters in place, relationality and affect, thus offers me an entrance point for talking about the lived experiences and responses of visitors to heritage places, including their affective responses to landscape, other visitors and heritage interpretation (Knudsen and Waade 2010). I explore these themes in Chapters 6 and 7, which deal with visitors' affective responses to their direct experiences of Aboriginal cultural heritage in place and how they believe themselves to have been changed by such experiences. I also utilise more-than-representational approaches in order to explain visitors' perception of authenticity in the tourist experience as something arising from their engagement with Traditional Custodian tour providers in the present moment (see Chapter 5).

The concept of affect, although arising within philosophical approaches prior to and separate from more-than-representational theory (Gregg and Seigworth 2010), figures prominently within this theoretical paradigm. Affect may be understood as a trans-personal and, some argue, pre-cognitive intensity capable of passing from person to person. As distinct from emotion, affect does not refer to a conscious feeling, but rather to a "pre-personal intensity corresponding to the passage from one experiential state of the body to another and implies an augmentation or diminution in that body's capacity to act" (Deleuze and Guattari 1987, xvi). While there are a range of different approaches to affect as set out by Melissa Gregg and Gregory Seigworth (2010), my application of the term is influenced by a vitalist philosophy, aligned with a Spinozian phenomenology, as articulated and expanded by Gilles Deleuze. This is a non-subject-based approach, but one that emphasises the body and materiality, in that it is concerned with "all kinds of things brought into relation with one another by many and various spaces through a continuous and largely involuntary process of encounter" (Thrift 2008, 8). However, I to some extent seek to recuperate a role for the human subject in the operation of affect, given my focus on the impact of heritage interpretation and touristic encounters upon the visitor through both cognitive (and non-cognitive) processes. I also argue that visitors' affective and emotional responses are informed by settler-colonialism and constitute what Mark Rifkin (2011, 350), has labelled "settler structures of feeling" and thus see the affective as linked to the discursive, rather than as wholly separate from it.

In adopting more-than-representational approaches as part of my conceptual framework, it is important to be cognisant of critiques of such approaches, including the concept of affect. These include questions about whether affect is, in fact, non-cognitive, and the supposed barriers that exist within the psyche between affect, emotion and cognition. I am persuaded by psychoanalytic approaches to affect that dispute the impermeability of these barriers, arguing that affective responses become knowable once they reach the level of emotion and cognition (Pile 2010). My position on affect is also informed by the arguments of theorists such as David Harvey (2015), who points out that the emphasis on the present moment and personal experience in phenomenological approaches runs the risk of reducing all that can be known to solipsism and presentism, neglecting the social, political and historical. Furthermore, I agree with Divya Tolia-Kelly (2006) and Deborah Thien's analysis (2005) that the human subject of non-representational theory is universalistic and based on a white male model, neglecting the specific relations in the world that may arise from bodies being black and female. Lorimer (2005) also points out that in its focus on the non-cognitive, affect is a theory about something that is essentially unknowable or only knowable in its processes/assemblages.

Additionally, it has been highlighted that non-representational theorists of affect such as Thrift are suspicious of emotion, which they view as "engineered by the powerful" (Pile 2010, 12). As Gary Campbell et al. (2017) note, a similar suspicion of emotion characterised earlier work in heritage studies, particularly attitudes towards nostalgia, which was regarded as a conservative, backward-looking glance to a fictitious past. In recent years, there has been a more complex theorisation of emotion within heritage studies to refer to more than conscious subjective feeling that is interior to a bounded human subject (Madgin and Lesh 2021; Marchant 2019; Tolia-Kelly, Waterton and Watson 2017). As Alicia Marchant (2019, 7–8) explains, "heritage is an accumulation formed in the relationship *between* the body, material cultures, rituals and emotion, recognising that the nature of this relationship is itself a dynamic process that changes over time, and is socially and culturally constructed" (emphasis in original). Similarly, theorists in the field of cultural geography, such as Liz Bondi et al. (2007, 3), refer to emotion as a socio-spatially mediated phenomena, understood as "relational flows, fluxes or currents, in-between people and places".

Like the flow of affect, the boundaries between people, places and things are thus perceived as permeable to emotion. In fact, Sara Ahmed (2004) argues that emotions are not simply "in" the subject or the object but come into being in relationships between people and things/objects. Rather than psychological states, Ahmed (2004, 9) describes emotions as "social and cultural practices" that are performed (for example, through speech acts) and involve investment in social norms. Similarly, Waterton (2014), Micieli-Voutsinas (2017) and other theorists of affect such as Tolia-Kelly et al. (2017), regard our affective responses as mediated by our socially and culturally situated bodies, and not outside of or separate from politics. I have found these insights helpful in interpreting my fieldwork experiences, particularly as a participant observer of cultural tours, and in my later discussions with visitors of their subjective perceptions of these tours. Understanding such encounters relies on a grasp of affect and emotion as inter-subjective, socially and culturally mediated and socio-spatial, as visitor responses were framed in their encounters with other visitors and cultural tour guides as they unfolded in space and time, but also informed by discourses about Aboriginality that circulate in the settler-colonial Australian context.

The desire to reclaim affect and emotion as vital to human and non-human co-constitutive becomings has arisen from a critique of the denigration of emotion within Western thought. It is also integral to the critique of discredited Cartesian ways of knowing implicated in the assumed inferiority of "nature", women and those peoples designated as "other" with colonial paradigms (Ahmed 2004). Cartesian logic is a fundamental organising component of a nature/culture dualism I argue is evident in the type of scientific thinking that has informed the management and interpretation of nature conservation areas by the settler-colonial state, (as discussed in Chapters 3). A reclaiming of the affective, emotional and embodied could thus be regarded as a form of decolonial theorising and an aspect of an Aboriginal onto-epistemology of relationality, that visitors must engage with if they are to connect with Country in any meaningful way (Kearney 2009; Bird-Rose 1996b; Wright et al. 2012). At the same time, I argue that it is important to consider how the different socio-political positionings and personal and collective histories of Aboriginal and non-Aboriginal peoples inevitably affect the ways we co-become together in the cultural tourism context.

Conclusion

There is clearly a degree of tension in my conceptual framework between the idea that subjects are created through discourse, which lends itself to the application of a postcolonial and a decolonial approach, and the idea that meaning is the product of momentary encounters between human and non-human others, in place and time, that is aligned with more-than-representational theory. By bringing these conceptual approaches together, I seek to demonstrate that they are not diametrically opposed but can inform each other, and that even momentary affective encounters are not devoid of social meaning. As I explore in Chapter 6, domestic visitors' emotional and affective responses to experiences of Aboriginal cultural heritage in the landscape, including feelings of pride, shame and anxiety about their role as tourists are typical of what Rifkin (2011, 350) has described as "settler structures of feeling". These feelings cannot be typified as random but rather are informed by visitors' encounters with Aboriginal peoples and places in a settler-colonial context, including visitors' awareness of Australia's history of colonial invasion and dispossession of Aboriginal and Torres Strait Islander peoples. This is why I argue that the discursive is entangled with the affective and emotional, and that although visitors may experience transformational moments, the status of many as privileged white subjects means that such transformations are incomplete. Rather than viewing representational and more-than-representational approaches as opposed, I seek to demonstrate their potential for productive engagement, leading to a deeper understanding of the cultural dimensions of tourism.

Notes

1 The term majority world is used to refer to the countries of the world where most people live, and the term minority world to refer less populated regions. As Punch and Tisdall (2012) explain, this terminology "... acknowledges that the 'majority' of population, poverty, land mass and lifestyles is located in the former, in Africa, Asia and Latin America, and thus seeks to shift the balance of our world views that frequently privilege 'western' and 'northern' populations and issues". This terminology is more in keeping with a decolonial approach.

2 Jäger and Maier (2016, 113) explain that "by dispositive (building on Foucault 1980a: 194), we mean a constantly evolving synthesis of knowledge that is built into linguistically performed practices (i.e. thinking, speaking, writing), non-linguistically performed practices (vulgo 'doing things') and materializations (i.e., natural and produced things)".

3 Sanctioned interpretations of culture and place

Discourses of protected area interpretation

Introduction

The previous chapter outlined the conceptual framework underpinning the research, including Foucault's theory of discourse and the way it is applied by contemporary scholars. In this chapter, I apply this approach to my own research, revealing how discourse can move from the written page into the landscape, including in the form of *in situ* interpretative signage. The chapter begins with a brief critical overview of relevant scholarship focusing on heritage interpretation within protected areas to support the discourse analysis that follows. It then moves on to analyse selected examples of heritage interpretation in each research locale, commencing with interpretive signage within Uluṟu-Kata Tjuṯa National Park. This signage is then compared with interpretive signage at Red Hands Cave in the Blue Mountains National Park. Next, two mobile phone tours on the Jenolan Caves app focusing on Gundungurra history and culture are introduced and discussed. The chapter concludes with a summation of the key findings of the critical discourse analysis and its relevance to the research as a whole.

Debates around protected area interpretation

Freeman Tilden's book *Interpreting Our Heritage*, originally published in 1957, was based on approaches to interpretation developed by the United States National Parks Service. Despite its focus on the United States, this book has been of great influence in defining what interpretation is, including across widely varying cultural and natural heritage locales (Slack 2021; Staiff 2014). Tilden (2007, 33) famously defined interpretation as "[a]n educational activity which aims to reveal meanings and relationships through the use of original objects, by firsthand experience, and by illustrative media, rather than simply to communicate factual information". Commenting on the influence of Tilden's work, Russell Staiff observes that because of Tilden's emphasis on education, this goal "was reinforced as the central characteristic of the interaction between visitors and heritage sites/places where interpretation activities were offered" (Staiff 2014, 19).

DOI: 10.4324/9781032654973-3

The primary goal of state-sponsored protected area education through interpretation was, and remains, to influence visitor attitudes and behaviours in favour of environmental conservation, by fostering their care and attachment to protected areas. Unfortunately, this focus can lead to the neglect of other meanings and values, including the meanings of landscapes to Indigenous peoples. This is particularly so when the dominant language used to convey interpretive messages in protected areas is that of Western science. As Staiff et al. (2002, 102) recognise, such a focus not only "communicates an assumption that this knowledge is universal, authoritative and, seemingly, outside the realm of contestation", but is also "organically bound to the landscape being experienced". A focus on Western scientific discourse within protected area interpretation fails to reflect the multiple and contested meanings of landscapes, marginalising Indigenous knowledge and other social, cultural and historic values associated with places.

A reliance on the epistemology of Western science within protected area interpretation ties in closely with Tilden's belief (2007, 33) that the role of interpretation is "the revelation of a larger truth" and that such truths are "universal". Such claims, although to some extent a product of their time, are also apparent in more recent statements by the managing authorities of protected areas about the type of narratives that should be told in these places. For example, the American National Park Service Advisory Board (ANPSAB) views parks as "cohesive, nationally distinctive landscape[s that] ... tell nationally important stories about our nation and are representative of the national experience" (ANPSAB 2006, 2). For the ANPSAB, parks are thus closely linked with national identity, belonging and beliefs about national cohesion (Gonzalez and Vazquez 2014). However, this view fails to take account that ideas of belonging by implication invoke those of unbelonging (Yuval-Davis 2006), and immediately construct an excluded "other" who exists within the gaps and silences of these national stories. Whether these be the Indigenous peoples historically excluded from living and hunting in protected areas (Kantor 2007), or other groups of people who have been found to underutilise parks, such as migrants and the low-income (Crang and Tolia-Kelly 2010), not everyone is included in these "nationally important stories".

Not only this, Staiff (2014, 99) argues that "the largely hidden (and often ideological) frameworks that are used in interpretation" often advocate what he calls a "progressive modernity", in which "history is the 'story' of continuously unfolding progress whereby the present/future is better than the past". Such a perspective is distinctly colonial and typically derogatory towards Indigenous lifeways, and thus particularly concerning when coupled with discourses that foreground Western science as a lens through which to interpret landscapes important to Indigenous peoples. Chance Finegan (2019, 284) thus argues that interpreters should be more critically reflective of their own practice, contending that,

> [i]nterpreters un/knowingly become complicit in ongoing colonial processes if they are working in an Indigenous context and do not critically

question the research on which they base their work and do not represent the multiple meanings that inhabit the resources they interpret.

Tilden's belief that the purpose of interpretation is to reveal universal truths rather than multiple meanings is also problematic because it assumes that places have settled, shared, pre-existing meanings. Also troubling is the assumption that an authority, in this case the heritage "expert", defines and communicates these truths to an audience that is simply a passive receptacle for this knowledge. Rather than revealing essential pre-existing values, Staiff (2014, 46) argues that heritage interpretation is better understood, after Stuart Hall et al. (2005), as a "system of representation", or, in other words, a set of discursive practices. As such, interpretation inevitably conveys social, cultural and political meanings, potentially in ways that reinforce institutional power (see also Graham et al. 2000). In fact, claims to universalism are often indicators that the perspective being reproduced is white, Eurocentric and male. This universalism is reflected in the tendency of Tilden and his contemporaries not to acknowledge that protected areas have ongoing significance to Indigenous peoples and that the history of their establishment was mired in a colonial order. This included imposing artificial national park boundaries with no regard to existing Indigenous relationships with landscapes and excluding Indigenous people from these newly created geographies. However, colonial practices such as these cannot simply be relegated to the past. The failure of settler nations such as Australia, New Zealand and the United States to recognise and prioritise the perspectives of Indigenous peoples in protected area interpretation is an ongoing colonial practice (Carr 2004; Finegan 2019; Hvenegaard et al. 2016; Staiff, Bushell and Kennedy 2002). This practice is not simply discursive but has the material affect of marginalising Indigenous peoples in landscapes that are important to them, as set out in Chapter 6, in the account provided of my discussion with three young Aboriginal women at Red Hands Cave in the Blue Mountains National Park.

Roy Ballantyne (1995) makes the important point that only Indigenous peoples, particularly the Traditional Custodians of a place, have the right to decide if their culture and history should be interpreted to outsiders, how this should be done, and by whom. Likewise, Finegan (2019, 285) questions the legitimacy of non-Indigenous guides interpreting Indigenous culture to visitors because "knowledge is culturally embedded" in the community in which it originates, "not something to be learned quickly and extracted from its context" (Finegan 2019, 289). Pointing to the importance of Indigenous peoples having control over their own cultural heritage, Ballantyne (1995) recognises that they have the right to withhold aspects of their culture from visitors by deciding that traditional knowledge associated with certain places not be interpreted, including through on-site signage. These observations about the rights of Indigenous peoples to control access to their cultures are in line with contemporary professional standards for interpreters in the Australian context and internationally and are consistent with international human rights standards.[1]

Early attempts by protected area management authorities to interpret Indigenous cultures resulted in their misrepresentation as ending with colonisation, rather than emphasising the continuing connections of Indigenous peoples with protected areas (Ballantyne 1995; Carr 2004; Staiff et al. 2002). Additionally, interpretation developed without the involvement of Aboriginal and Torres Strait Islander peoples has not lent itself to "truth-speaking", because it has tended to exclude "controversial issues such as the Stolen Generations, mission stations, and massacre sites" (Ballantyne et al. 2012, 155). However, more recent site-based interpretation such as that in Uluṟu-Kata Tjuṯa National Park, as Clarke and Waterton (2015) observe, also excludes accounts of colonial violence and the park's more contemporary contested history, for example, the opposition of the NT Government to the park's 1985 return to Aṉangu.[2] The types of discursive absences identified above, plus the continuing appearance of Western scientific narratives that perpetuate a nature/culture dualism, are elaborated below in a discourse analysis of interpretive signage in Uluṟu-Kata Tjuṯa National Park and the Blue Mountains National Park, followed by a discussion of digital interpretation offered on the Jenolan Caves mobile phone app.

Interpretation in Uluṟu-Kata Tjuṯa National Park

There are three main ways in which Parks Australia, in partnership with Aṉangu, offers interpretation to visitors to Uluṟu-Kata Tjuṯa National Park. These are through the Uluṟu Cultural Centre, interpretive signage located at key locations around the park and a free ranger guided walk along the short Mala walk at the base of Uluṟu, offered daily. I discuss visitor responses to interpretive signage, the ranger guided walk and the Uluṟu Cultural Centre in Chapters 5–7, focusing in this chapter on a discourse analysis of interpretive signage and digital forms of interpretation in the landscape. As space does not permit discussion of all the interpretive signage in the park, I have chosen to focus on specific examples of signage at two locations considered representative of other signage in the park. These are the signage located adjacent to the main road into Uluṟu-Kata Tjuṯa National Park, at a parking bay Parks Australia refers to as the "Car Sunset Viewing Area" and selected signage along the walks at the base of Uluṟu. According to Parks Australia, the current signage at these locations dates from 2010, and Aṉangu were highly involved in its development.

Discourse positions

The Palya, or welcome sign (Figure 3.1), is located at the "Car Sunset Viewing Area" along Uluṟu Road, close to the park entrance. It is typically the first interpretive sign viewed by visitors when they stop to take a panoramic photo of the Uluṟu monolith in the distance. Under the heading Palya, the first words of the sign state "Aṉangu traditional owners welcome you to

Figure 3.1 "Palya-A̲nangu traditional owners welcome you to A̲nangu land" Interpretative Sign. Photograph by the author.

A̲nangu land". While ostensibly written to welcome the visitor, the sign also immediately establishes an A̲nangu claim to sovereignty over the land that is now Uluru̲-Kata Tjut̲a National Park, setting the tone for the interpretive signage that follows. Utilisation of the first person A̲nangu voice in this signage signals that it is not in the same mode as earlier signage, discussed above, that represented Indigenous peoples as absent from contemporary protected area landscapes (Ballantyne 1995; Carr 2004; Staiff et al. 2002). The interpretive text shown in Figure 3.1 has multiple discourse positions, a quality Jäger and Maier (2016) associate with non-hegemonic texts. There are two dominant discourse positions, that of A̲nangu and Parks Australia, and a shared discourse position between A̲nangu and Parks Australia that appears intermittently, for example in the opening statement under "Both Ways" and the closing statement located directly above the line drawing of Uluru̲. This shared discourse position is signalled by the first person plural pronoun "we" in the following excerpt from the text in Figure 3.1 under "Both Ways",

> A̲nangu and Parks Australia jointly manage the park. *Traditional land management and western science* work together to preserve the park's *outstanding values* under the guiding principle of Tjukurpa ... See how *we* view this land from these two perspectives, learning from each other. (Emphasis added)

The text under the heading "Both Ways" positions Anangu and Parks Australia as working together, using "Traditional land management" and "western science", presenting these as two distinct but complementary approaches employed in managing the park. The two voices speak together as "we" under the heading of "Both Ways"; however, the formal language and tone of this section and the reference to the World Heritage discourse of the park's "outstanding values" are more representative of the "expert" voice of Parks Australia, than the discourse position of Anangu. The shared discourse position is intended to convey the unity of Anangu and Parks Australia in managing the park, a discourse of harmony and cooperation that elides any discord, fractures and structural inequalities in the relationship, however, it is one in which formal "expert" language dominates.

A singular Parks Australia voice rather than a shared voice is apparent in Figure 3.1 on the left of the panel, speaking about the landscape using natural resource management (NRM), heritage and tourism discourses. This text is divided into sections using the heritage terms "Natural Values" and "Cultural Values" as headings. Under "Natural Values", the park is described as "famous for its *exceptional natural beauty*" (emphasis added), framing it as an aestheticised landscape suitable for viewing by tourists. The text under the heading of "Cultural Values" emphasises the landscape's "religious and cultural connections" to Anangu. However, it also states that this "ancient land has been actively *managed* by generations of Anangu", describing Anangu's relationship with this landscape using the language of natural resource management rather than the Aboriginal ontology of Country. The primary ideological position communicated in this text is thus the Eurocentric ontology of Parks Australia that understands the park as a touristed landscape to be "managed" to preserve its natural and cultural values, with its "religious and cultural connections" to Anangu treated as subsidiary. NRM discourses thus effectively marginalise an Anangu relationship with Country in site-based interpretive signage within the park (Howitt et al. 2013).

The third discourse position appearing in Figure 3.1, on the far right under the heading "Mai Kalka (edible seed)", is that of Anangu women, although their discourse is accorded less space proportionately than that of Parks Australia. This section of the panel explains how Anangu women collect grass seed to make flatbread, stating in the first-person plural, "*We* run our hands along the seed heads and scoop them into a piti (large bowl). *We* clean the seed, grind it and mix it with kapi (water)" (emphasis added). The use of Pitjantjatjara in the heading and for other common nouns that identify plants and objects used in bread-making, such as "piti", signals that the Anangu voice is speaking, as does the use of the first-person plural pronoun "we" to refer to Anangu women. This is an active, embodied voice, engaged in processes of living, including taking special care of children by sometimes adding "tjala (honey ant)" to sweeten the bread, before adding it to the coals. The story of bread making thus evokes an intimate picture of family life, easily relatable to visitors, and the use of the present tense positions this practice as part of a living, continuing culture.

The voice of A<u>n</u>angu is most apparent in Figure 3.1 when directly addressing the reader in a mixture of Pitjantjatjara and English and is typified by emotive, embodied, informal and often colloquial language. The marginalisation of A<u>n</u>angu ontologies through the NRM and heritage discourses discussed above, is thus to some extent countered through the embodied and relational A<u>n</u>angu voice expressed in this text fragment. This apparent contradiction is better understood as a presentation of multiple perspectives, a characteristic Finegan (2019) and Hvenegaard et al. (2016) identify as an appropriate contemporary interpretative strategy for protected areas, and one that is also now common in museums (Watson 2009).

Although the first words of the sign "A<u>n</u>angu traditional owners welcome you to A<u>n</u>angu land" immediately conveys the primacy and authority of A<u>n</u>angu in this place, the semiotics of the sign's location construct Ulu<u>r</u>u as an object configured for the tourist gaze, echoing the iconic panoramic view of Ulu<u>r</u>u first captured photographically by Walter Baldwin Spencer in 1894 (Wallis 2014), and since then replicated in thousands of tourist images. The area Parks Australia describes as the "Ulu<u>r</u>u Car Sunset Area" is located some distance from Ulu<u>r</u>u, and enables it to be viewed panoramically, particularly as the colour of Ulu<u>r</u>u changes at sunset. The viewing area becomes very crowded at this time of day as visitors congregate to take photographs. The location and orientation of this sign encourage the framing of Ulu<u>r</u>u as an aesthetic landscape feature that is an object of the visitor's gaze, a form of visual framing that reflects a scopic, colonising approach to place (Urry 2011). The use of the Western convention of line drawing in the illustration of the landscape provided in Figure 3.1, and in other interpretive signs in the park discussed in this section, visually presents the landscape for viewing by a Western audience, using familiar visual cues such as perspective.

Rather than facilitating an understanding of the whole landscape as having meaning to A<u>n</u>angu, the focus on Ulu<u>r</u>u achieved through the placement of the welcome sign narrows the visitor's gaze to this one feature, reflecting the settler-colonial obsession with Ulu<u>r</u>u since first viewing this landscape (Barnes 2011; Hueneke and Baker 2009; Wallis 2014). By inviting visitors to visually claim this landscape, utilising perspectival ways of seeing not only in the gaze invited by the sign's physical placement but in its line drawing illustration, the message of the sign that the park is A<u>n</u>angu land is to some degree undermined. The interpretive signage discussed in this chapter can thus be characterised as demonstrating a type of doubleness or ambiguity, in which the ostensible priority given to A<u>n</u>angu perspectives is countered by a continued focus on Western scientific discourse and Western aesthetic conventions. Additionally, the physical presence and configuration of the interpretive signage itself literally marks the park as a touristed landscape. Regardless of the content of the signage, its simple presence could be understood as a form of

Figure 3.2 Kuniya Walk Interpretative Sign, UKTNP. Photograph by the author.

"deep colonising", after Bird-Rose, who uses the term to refer to "conquest embedded within institutions and practices which are [ostensibly] aimed toward reversing the effects of colonisation" (Bird Rose 1996a, 6).

Unlike the Palya sign, which expresses a variety of discourse positions, the sole voice speaking in the sign, Kuniya Walk (Figure 3.2), is that of A<u>n</u>angu. As the voice is genderless, we can assume that both A<u>n</u>angu men and women are speaking. This voice is authoritative and direct, but also informal and relatable in the use of colloquial language such as "have a think" in the excerpt below, instructing the visitor to:

> Look at the landscape as we do and know these ancestors *are still here*. This is the right place to learn about this story because it happened here at this place. Look, take note of what you see. Stop and read the signs, *have a think* and take in your surroundings. This is a place of *great history*, an important place. (Emphasis added)

The speaker is one who has a claim to this place because they know about its history and the Tjukurpa associated with it. Expressing an A<u>n</u>angu onto-epistemology, the speaker states unequivocally that the ancestors associated with this Tjukurpa "are still here", and invites visitors to connect with it as an important place of "great history". The speaker is sharing this knowledge with the viewer, who, it is assumed, is not part of the A<u>n</u>angu "we" who possesses this knowledge, but someone other, who needs to be instructed on how to see the ancestors in the landscape and behave in this place. A<u>n</u>angu frequently adopt a pedagogical discourse position in the signage at the base of Uluṟu in line with their responsibilities as Nguraṟitja (Traditional Custodians) to instruct visitors about the meaning of places and the appropriate way to behave in them. A<u>n</u>angu thus demonstrate a degree of ownership of this interpretation, as they adapt it to their purposes, distinguishing Uluṟu-based interpretative signage from earlier examples of interpretation in protected areas that exclude Traditional Custodian voices and perspectives (Ballantyne 1995; Carr 2004; Huggins et al. 1995; Staiff et al. 2002).

Discourse strands

Continuity of a living A<u>n</u>angu culture

A key discourse strand emerging from the set of interpretive signage under discussion is that A<u>n</u>angu culture is strong and living, in direct continuity with its own past. For example, in the "Mai Kalka" text in Figure 3.1, the activity of making bread using grass seed is situated in the present through present tense verbs such as "run" and "clean". Positioning the park as a place that "continues to hold powerful religious and cultural connections" in the same sign also conveys the living dynamism of A<u>n</u>angu culture. Such language avoids locating A<u>n</u>angu beliefs and practices such as bread-making as relics of the past, as some older-style interpretation in other parts of Australia and New Zealand position Indigenous lifeways (see discussion of Red Hands Cave, NSW).

However, the interpretation does not explain the way that practices such as bread-making have changed in the last 50 or so years, particularly since store-bought bread became locally available. During the Pu<u>n</u>u Puṯitja – Bush Plants tour at the Uluṟu Cultural Centre that I participated in May 2019, I learnt that gathering grass seed to make the bread referred to in Figure 3.1 is time-consuming, and now tends to be done only on special occasions. Failure to explain the specific ways in which traditional practices for obtaining food have now changed, including the use of motorbikes to hunt game, is a feature of interpretive signage within Uluṟu-Kata Tjuṯa National Park. Although the description of traditional food production practices such as bread-making expresses cultural continuity, it runs the risk of reinforcing an authenticity discourse that romanticises A<u>n</u>angu culture as timeless and unchanging, a familiar trope in protected area interpretation of Aboriginal cultural heritage identified by Clarke and Waterton (2015).

However, it must be acknowledged that locale-based interpretive signage is not ideal for communicating complex or nuanced meanings, since this medium requires brevity. This is because of the short periods of attention most visitors give to reading it and the need to restrict wordage for ease of comprehension (Bryant 2006; Cole et al. 1997; Hughes and Morrison-Saunders 2002). As a consequence, visitors who participate in Traditional Custodian led cultural tours may gain a deeper understanding of changing

A<u>n</u>angu lifeways, than those who rely exclusively on interpretive signage (see Chapter 4).

Fostering an embodied connection with Country

The sign headed Kapi (Figure 3.3), named after the Pitjantjatjara word for water, is located at Kantju Gorge at the base of Ulu<u>r</u>u. It is the most explicit direction given in the signage discussed in this chapter to connect with Country through the senses, asking "Can you hear the wind? Can you feel the temperature difference? Take a minute to sit quietly, close your eyes and breathe deeply. Feel this special place, enjoy this moment". Similarly, the Kantju Gorge sign (Figure 3.4) encourages the non-A<u>n</u>angu viewer to carefully approach the Gorge in a thoughtful and attentive manner. The call to be attentive, not just to the voice of A<u>n</u>angu speaking through this signage, but to one's surroundings, using not only the senses but also the emotions, is typical of the tone and content of much of the onsite signage around Ulu<u>r</u>u. This is made most explicit in Figure 3.3's exhortation to "Feel country" by attuning to the sense of hearing and bodily sensations of temperature. An invitation by A<u>n</u>angu for visitors to connect with Country is apparent in the Kapi sign and other signage in the park such as at the Mutitjulu waterhole. Such messaging opens up the Aboriginal ontology of Country to visitors by inviting an embodied relationality with Country, even though their understanding of and capacity to engage with Country, as non-A<u>n</u>angu, is necessarily limited (see Chapter 7). This embodied relationality is a feature not

Figure 3.3 Kapi Interpretative Sign, UKTNP. Photograph by the author.

Figure 3.4 Kantju Gorge Interpretative Sign, UKTNP. Photograph by the author.

only of an Aboriginal ontology (Brigg and Graham 2020; Tynan 2021) but features prominently in Indigenous-led cultural tourism experiences (Bawaka Country et al. 2017; Beck and Somerville 2002).

Positioning the park as an A̲nangu place and the primacy of Tjukurpa

The prominent positioning of the Pitjantjatjara word Palya in Figure 3.1, followed by the statement, "A̲nangu traditional owners welcome you to A̲nangu land", identifies the park as an A̲nangu place, a key discourse strand running through this signage. This sign also introduces the goal of "preserve[ing] the park's outstanding values under the guiding principle of Tjukurpa- the foundation of all A̲nangu knowledge". Tjukurpa is presented as an overarching principle that guides the work of the park, consistent with the way it is positioned in the park's 2010 Management Plan. The importance of Tjukurpa is also apparent in Figure 3.2, Kuniya Walk, in the A̲nangu statement that "Tjukurpa is the basis of all our knowledge, law, religion, social structure and moral values".

Much of this signage also attempts to establish behavioural norms for visitors based in Tjukurpa, such as continuing to treat the Mutitjulu waterhole as "a quiet place of respect" (Figure 3.2, Kuniya walk). Similarly, Kantju Gorge (Figure 3.4) begins with the admonition "Please do not swim or disturb this special place in any way", explaining "We learn to look after and respect these places because they are the beginnings of Tjukurpa". The visitor is again asked to show respect for Tjukurpa by allowing it to dictate their behaviour in this place, with the A<u>n</u>angu speaker performing their responsibilities as Ngura<u>r</u>itja (Traditional Custodian) by conveying this message. Although I argued earlier in this chapter that the presence of interpretive signage in the landscape of the park could be regarded as a form of "deep colonising" (Bird Rose 1996a, 6), A<u>n</u>angu also Indigenise this medium, to some degree taking control of it to fulfil their responsibilities as Ngura<u>r</u>itja. I position this as a form of Indigenisation, rather than decolonisation, since it occurs in the context of the use of this land as a national park and a site for tourism (Banivanua Mar 2010), uses over which A<u>n</u>angu lack control, despite their formal ownership of the land that comprises the park.

Heritage discourses

Despite the prominence accorded to an A<u>n</u>angu worldview within the signage under discussion, heritage discourses that position the park as a World Heritage Area, which is "a special place … recognised by UNESCO as … of *outstanding universal value*" (emphasis added, Figure 3.1, Palya Sign) for both its natural and cultural features also figure prominently. As discussed above, the discourse position in this section of text in Figure 3.1 is that of a government heritage "expert", since terms are utilised that only those familiar with the language of heritage would understand, such as "outstanding universal value". The concept of outstanding universal value that appears within this expert commentary is based on the principle that this landscape, as World Heritage, is now the patrimony of all the world's peoples, which diminishes the A<u>n</u>angu claim as merely one among many. This way of thinking is not in keeping with Tjukurpa, which establishes A<u>n</u>angu law and meanings of place as primary.

Furthermore, Figure 3.1's description of Ulu<u>r</u>u as an "ancient" landscape reflects a Parks Australia discourse position that is more in keeping with a Eurocentric understanding of time, rather than A<u>n</u>angu notions of sacred time, in which Tjukurpa occurs not only in the past but also in the present (James 2015). An A<u>n</u>angu discourse position is more accurately conveyed in Figure 3.4, Kantju Gorge, in the text that explains that not only "generations and generations" of A<u>n</u>angu but also "wildlife" relied on the water in this gorge to survive. This text is more attuned to Aboriginal ontologies of relationality than Western notions of time-depth, as it expresses the past in terms of human and non-human connections with Country rather than sums of years (Dudgeon and Bray 2019; Tynan 2021).

I argue that an emphasis on the time-depth of A<u>n</u>angu culture is a substrand within heritage discourses of Aboriginality and also a component of a settler-colonial discursive formation that articulates Aboriginality in terms of discourses of authenticity and inauthenticity. Within this discourse, Aboriginal peoples who are assessed as "authentic" exist in a continuing relationship with an ancient, unchanging, monoculture (Attwood 1992; Cowlishaw 2010, 2011; Lattas 1992), whereas those who are judged inauthentic are not able to demonstrate these relationships (see Chapter 5). Settler-colonials, it has been argued, are drawn to this idea of an ancient "authentic" Aboriginality "in a nationalist search for a deep past European Australians ... feel they have never had, and a spirituality they have apparently lost", as a way of legitimising their claim to Australia (Attwood 1992, iv; see also Byrne 1996).

A nature/culture dualism

A common feature of the authenticity discourse outlined above is a nature/culture dualism that locates an "authentic" Aboriginality in close proximity to a pre-existing "nature", for example, in a protected area setting. A nature/culture dualism is apparent in the signage under discussion in the contrasting style and tone of text expressing an Anangu and a Parks Australia discourse position. While an embodied, emotive Anangu voice exhorts the visitor to connect with the landscape, the formal, "expert" voice of Parks Australia interprets the landscape from the perspective of Western science. Such dualities reinscribe a familiar Cartesian mind/body split, particularly when the signage makes this binary explicit by inviting visitors to "continue your journey with your *heart* and *mind* open to experience this place *both ways*" (emphasis added, Figure 3.1 Palya sign).

The dichotomy between Western Science and Anangu traditional knowledge is a further way in which a nature/culture dualism expresses itself in the park's interpretive signage. As discussed above, Figure 3.1 establishes a "both ways" approach to park management, in which Parks Australia is associated with the dominant term, "Western Science", and Anangu with the lower status term in this dyad, "Traditional land management". While the interpretive signage attempts to position both perspectives as equally important, this is undermined by the weight of meaning already attached to these terms within hegemonic Eurocentric knowledge systems.

Byrne et al. (2013, 1) describe "the culture-nature divide as an ontological marker of Western modernity", and one in which "modern Western thought, society (or culture) has been understood to be not just radically separate from nature but situated in a hierarchical relationship of dominance to it". Haila (2000) characterises the nature/culture dualism as one of the most pervasive dualisms of Western metaphysics, a way of thinking heavily implicated in environmental degradation, as it poses humans as both separate from and superior to nature. A nature/culture dualism is also deeply colonial in that it contrasts a modernist Enlightenment rationality with the ways of being of peoples labelled "traditional" and closer to nature, whose beliefs are dismissed as myth and superstition (Quijano 2007).

The precedence accorded to Western scientific knowledge over Aboriginal traditional knowledge within colonial paradigms is in line with Foucault's (2002) understanding of the way that certain discourses are given the status of "truth" and regarded as "knowledge", while others are excluded from this category. Truth claims based on a nature/culture dualism underlie what

Wearing and Huyskens (2001, 189) describe as Australia's "Eurocentric approaches in national park policy". Writing of the impact of a nature/culture dualism in protected areas upon Indigenous peoples, trawlwulwuy woman Emma Lee (2016, 356), describes it as "a privileged and constructed framework that simultaneously undermines Indigenous power and authority in order to manage protected areas whilst enhancing the status of natural values". Despite the sustained scholarly critique of nature/culture dualisms and attempts to address these within approaches to managing and interpreting Australian protected areas (Brown 2021; Harrison 2015; Bird Rose 2003; Bird Rose et al. 2003), this way of thinking has proven to be deeply embedded and persistent, evident in relatively recent interpretation such as that found in Uluṟu-Kata Tjuṯa National Park.

Prohibited discourses

Discourses at odds with romantic narratives of the sustainability of Aṉangu lifeways tend to be omitted from interpretive signage within the park. One noteworthy absence of this type is any reference to the colonial or more contemporary history of the park, including at sites where important events took place during this era. The colonial period did not commence until the 1870s in this part of Central Australia and dates until at least the 1940s and is closely enmeshed with the history of the land that now constitutes the park. This history includes, but is not limited to, the actions of police and pastoralists, and the impact of these on Aṉangu and other Aboriginal peoples of Central Australia (Clarke and Waterton 2015; Edwards 2011; Hinkson 2018). The inclusion of a brief history of colonial/settler pastoralism in park-based interpretation would conform with the park's own natural resource management discourse, as habitat degradation resulting from the introduction of hooved livestock to this region is associated with significant animal extinctions. A focus on the history of pastoralism in Central Australia would also demonstrate the long struggle for justice and equality of the Aboriginal peoples of this region, beginning with the strike by Aboriginal stockmen and domestic staff at Wave Hill Station, NT in 1966 known as the Wave Hill walk-off (Charola and Meakins 2016). This history would tie in well with that of the Aṉangu struggle for land rights, which ultimately led the Australian Government to return Uluṟu-Kata Tjuṯa National Park to Aṉangu in 1985. Although the story of "hand-back" is provided in a video within the Uluṟu Cultural Centre, this important history should also be told within site-based interpretation, particularly at the location of the 1985 "hand-back" ceremony.

Turning to the history of interaction between police and Aṉangu, frontier violence had a direct bearing on the history of the park since it led to the shooting death of a member of the Uluṟu family by a police officer at Uluṟu in the 1930s. This event, although of considerable importance to the prominent Uluṟu family (Cowley 2018), is mentioned only briefly within interpretation at the Uluṟu Cultural Centre, despite repeated requests by A<u>n</u>angu that it receive greater attention, as reported by a Parks Australia staff member during a 2019 interview. At the time of my 2019 fieldwork, there was no on-site signage describing this event or identifying where it occurred. Accounts of this type would help fill the discourse gap around the colonial history of the land that now constitutes the park and perhaps enable visitors to make connections with ongoing social and political issues that impact Aboriginal peoples in settler-colonial Australia, such as their continuing deaths at the hands of systems of carceral control (Johnston 1991). The absence of any direct reference to the post-contact history of the park in on-site interpretive signage represents a discourse strand that is conspicuous by its absence, suggesting the existence of a prohibited discourse. This is an important gap not only within Uluṟu-based interpretation but also within on-site interpretation at Red Hands Cave in the Blue Mountains National Park, discussed below.

Interpretation of Red Hands Cave, Blue Mountains National Park

Red Hands Cave is what is termed in archaeological parlance an "Aboriginal rock art site", because of the coloured hand prints and stencils and other cultural markings within this rock shelf, which signage at the cave identifies as at least 1600 years old (see Figures 3.5 and 3.10). What follows is a critical discourse analysis of interpretive signage associated with Red Hands Cave

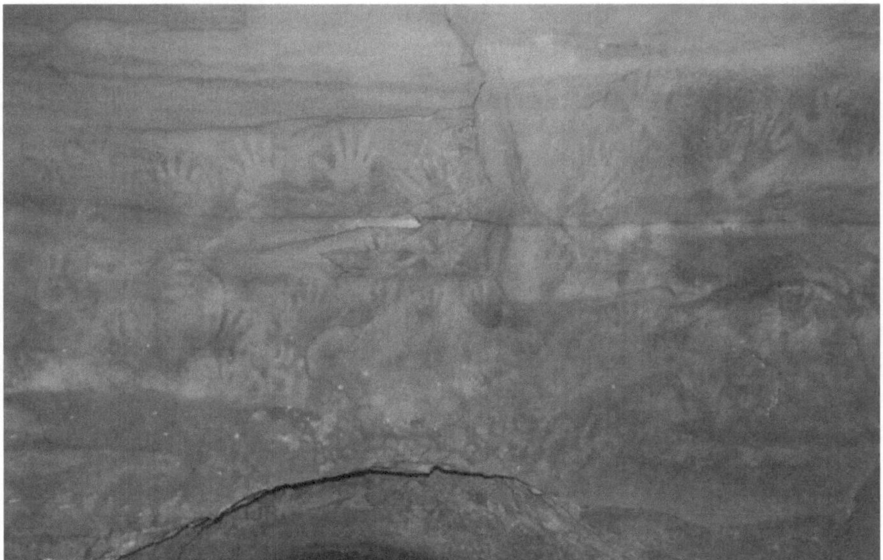

Figure 3.5 Hand prints and stencils present in Red Hands Cave, BMNP. Photograph by the author.

to highlight its discourse position/s, discourse strands, prohibited discourses and any emerging discursive formations. Space does not permit discussion of each sign in detail; rather, key discursive features are identified that are considered representative of the signage as a whole and presented thematically. The signage that I discuss is located within the cave, at the car park for the cave and along the walking route that leads to the cave. It was developed in 1988 through grant funding obtained by the NSW National Parks and Wildlife Service (NSW NPWS) (Cameron 2014). Consultations with staff of the NSW NPWS suggest that the signage has not been updated since its installation and that it was developed without the involvement of Blue Mountains' Traditional Custodians or other Aboriginal peoples.

Unlike Uluṟu-Kata Tjuṯa National Park, no regular tours guided by park rangers are offered to Red Hands Cave, although an NSW NPWS guide can be arranged on a user-pays basis, usually for school groups. There is also no information centre where visitors can speak to an NSW NPWS staff member, only a staff member at the entrance gate where visitors driving into the national park pay for ticketed entry. The greater attention to the interpretation of Aboriginal cultural heritage within Uluṟu-Kata Tjuṯa National Park demonstrated by a staffed Cultural Centre open to the public, well-maintained and regularly updated interpretive signage and free park-sponsored activities such as the ranger guided walk reflects its higher profile as an Indigenous tourism destination. This higher profile may be related in part to the significance of Uluṟu-Kata Tjuṯa National Park, in comparison with the Blue Mountains National Park, as a destination for international visitors, who show more interest than domestic visitors in Indigenous tourism (Mahadevan 2018). These different visitation patterns notwithstanding, I argue that the greater attention paid to Indigenous tourism in Uluṟu-Kata Tjuṯa National Park is also indicative of a settler-colonial assumption that "authentic" Aboriginal heritage deserving of well-developed and maintained interpretation is located in remote parts of Australia rather than on the fringe of a major city such as Sydney.

Discourse position

The interpretive signage related to Red Hands Cave reveals a unitary discourse position, a characteristic Jäger and Maier (2016) associate with hegemonic narratives. This single perspective is that of the white "expert" speaking on behalf of the NSW NPWS, providing a generally distant scientific commentary typical of a natural resource management and conservation agency. This signage does not reflect current best practice interpretive standards that favour a "multiple voices approach" (Hvenegaard and Lemelin 2016, 47), including within a protected area setting (Staiff et al. 2002).

In contrast to the interpretive signage at Uluṟu-Kata Tjuṯa National Park, discussed above, there is a complete absence of a Traditional Custodian voice or a shared discourse position in interpretive signage associated with Red

Hands Cave. The dominant white discourse position is apparent in the repeated use of the third person pronoun "they" and "their" to refer to the Dharug people and the use of the word "us" to refer to the audience of the signage, who are assumed to be non-Aboriginal. The whiteness of the "us" and "we" is made explicit in the sentence "[i]f *we* were left to survive in this bushland *we* would probably die" (emphasis added, Figure 3.8, The Neighbourhood), part of a narrative about the ability of the Dharug people to sustain themselves in an environment in which white settlers could not.

The sign headed "The Neighbourhood" appears on the walking route to the cave and seems intended to convey the familiarity of the Dharug people with this environment, reflected in the patterns of utilisation that the sign describes. However, the term "The Neighbourhood" is also what Jäger and Maier (2016) identify as a collective symbol, one that will be understood by the assumed white audience, as it is a modern Australian colloquialism referring to a populated environment close to a person's home, evoking a possessive sense of belonging (Hage 2000). The use of this term serves to reorientate this place to a settler audience while simultaneously distancing the bushland environment that the sign refers to from Aboriginal ontologies that frame this place as Country. Such language represents one way that older-style protected area interpretation literally rewrites "the spatiality of Indigenous lands" (Banivanua Mar 2010, 75–6), recreating it as a place orientated to settler-colonial visitors.

This is a discourse in which a white "expert" is speaking about the Traditional Custodians of this place to other white people, in the Traditional Custodians' absence. The discourse is, in fact, predicated on their absence, signalled by repeatedly framing their occupation of the area in the past tense. This discourse position demonstrates a way of representing Aboriginal peoples that Hodge and Mishra (1990, 27) describe as "Aboriginalism", as it portrays Aboriginal peoples as the silenced "other", the subaltern who cannot speak (Spivak 1988). Hodge and Mishra (1990, 27) developed the term "Aboriginalism" to point to the similarities between settler-colonial constructions of Aboriginality and Edward Said's 1978 concept of "Orientalism" (see Chapter 2). They describe "Aboriginalism" as a "double movement, a fascination with the culture of the colonised, along with a suppression of their capacity to speak" (Hodge and Mishra 1990, 27). Such commentary is another way in which interpretive signage enacts a form of "deep colonising" (Bird Rose 1996a, 6), because it purports to respect and value Aboriginal cultural heritage, while appropriating and misrepresenting it, by excluding the first-person voices of Aboriginal peoples.

The discourse position that is conspicuous by its absence is that of the Traditional Custodian speaking from their own standpoint about their culture. This is particularly apparent in the failure to utilise the Dharug and Gundungurra languages within the signage. However, a Dharug Traditional Custodian consulted for this research questioned the appropriateness of employing a Traditional Custodian voice in this context, explaining that the decision to open this place to the public was not made by Aboriginal peoples and that the park remains Crown Land that is neither owned nor managed by them. As he argues in relation to the absence of a Traditional Custodian voice in the signage at Red Hands Cave,

> … part of me sits comfortably with that because I like people to understand, 'Yes, this is not us, your being allowed here is not by our doing. You're walking up these steps. We did not make these steps'. So my concern might be it looks a little bit too much like we're being, 'It's all okay', if it's addressed from us.

The postcolonising discourse position that underlies this text thus, in his view, reflects the current context in which it is presented, including the lack of control that Dharug Traditional Custodians of the Blue Mountains have over access to and interpretation of places of significance to them in the Blue Mountains National Park.

Such views highlight the importance of developing interpretation of places like Red Hands Cave only after careful and thorough consultation with Traditional Custodians and other Aboriginal peoples residing in an area. As part of this consultative process, agreement firstly needs to be obtained about whether the public should have access to such places, whether interpretation should be provided, and what form this interpretation should take (see Chapter 8). I assumed when I first conceived the idea for this research that updating the language of this signage and inserting a Traditional Custodian perspective were the only correctives required; however, after speaking with the Traditional Custodian quoted above, I came to understand that doing so would perpetuate the disempowerment this Dharug man describes and represent another form of "deep colonising".

Discourse strands

Extinguishment of Aboriginal peoples and cultures

One of the key discourse strands in this set of interpretation is that the Aboriginal occupation of Red Hands Cave and its surrounds is confined to the past, giving the mistaken impression that Dharug and Gundungurra people no longer live in the Blue Mountains or have ties to this place.[3] This signage is typical of older-style protected area interpretative signage that depicts the relationships of Indigenous peoples to protected areas as ending with colonisation (Ballantyne 1995; Carr 2004; Clarke and Waterton 2015; Porter 2010; Staiff et al. 2002). This is evident in the repeated use of the past tense to describe the Dharug people, who are sometimes simply referred to as "Aborigines". The use of the term "Aborigines" further signals the outdated nature of this signage, as it refers to Aboriginal peoples as a monolithic group rather than recognising their diversity. For example, the first

Figure 3.6 Welcome to Red Hands Cave Interpretative Sign, start of Red Hands Cave walk, BMNP. Photograph by the author.

sentence of Figures 3.6 and 3.7 reads "[f]or at least 14,000 years *Aborigines* (sic) *lived* in the Blue Mountains" (emphasis added). Various scholarly guides to appropriate contemporary terminology for referring to Aboriginal and Torres Strait Islander peoples exclude the term "Aborigine" from the category of respectful and appropriate language (QUT 2021). The use of

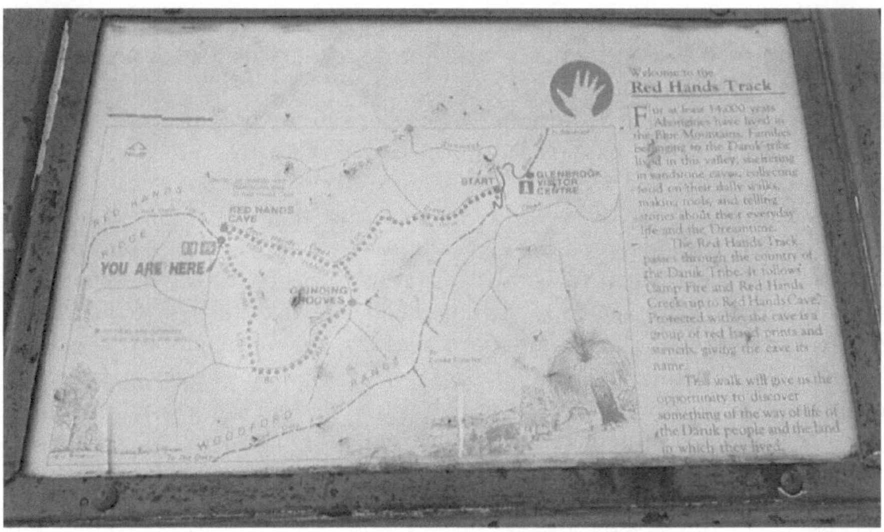

Figure 3.7 Identical welcome sign to Figure 3.6, Red Hands Cave car park, BMNP. Photograph by the author.

Figure 3.8 The Neighbourhood Interpretative Sign (located midway along walk to Red Hands Cave), BMNP. Photograph by the author.

the past tense is again evident in Figure 3.8, The Neighbourhood, which observes that this "land *was* able to sustain healthy Daruk[4] families for over ten thousand years" and in the signage within the cave which claims that ochre, "*was considered* one of the Aborigines' (sic) most prized possessions" (emphasis added). Ochre is, in fact, still used by Aboriginal peoples of the Blue Mountains and Yellomundee, as demonstrated during the two local cultural experiences with Aboriginal guides I participated in, discussed in Chapter 4. In these tours, ochre and its uses were introduced to visitors as an ongoing Dharug cultural practice.

While the absence of Traditional Custodians in the present day is implied through the use of the past tense, this absence is directly stated in the regulatory sign within Red Hands Cave (Figure 3.9), in the context of asking visitors to help protect "a way of life which is *now past*" (emphasis added). By stating this without emphasising Traditional Custodians' ongoing connection to this place, the text reinscribes the racist myth of Aboriginal people as "a dying race" (Armitage 1995, 18), embedded in the policies of Australian governments towards Aboriginal peoples in the protectionist era (i.e., 1860–1930, according to Armitage 1995, 27). While this policy approach is long abandoned, this signage suggests that the discourses associated with it continue to influence mainstream cultural expression decades after, with the idea of Aboriginality operating as a de-historicised floating signifier, gaining its meaning not from any racial essence, "but in the shifting relations of difference ... with other concepts and ideas in a signifying field" (Hall 2021, 362). The living presence of Aboriginal peoples is erased from the Blue Mountains through this discursive positioning, enabling non-Aboriginal

Figure 3.9 Regulatory Sign within Red Hands Cave, BMNP. Photograph by the author.

people to lay claim to this place, a form of elimination that facilitates white possession (Moreton-Robinson 2015; Wolfe 2006).

The use of the term "Aboriginal relic" in the regulatory sign within the cave (see Figure 3.9) to describe the material culture of Aboriginal peoples is also redolent of the idea that Aboriginal cultures are dead rather than living. Literal definitions of the term include "a survivor or remnant left after decay, disintegration, or disappearance" and "a trace of some past or outmoded practice, custom, or belief".[5] The term, when applied to Aboriginal cultural heritage, firmly locates it in the past, with the implication that places such as Red Hands Cave no longer have continuing relevance to living Aboriginal peoples. That these pronouncements are made in the name of the Director of National Parks on behalf of the NSW Government, makes them extremely authoritative, giving them the status of fact or truth (Foucault 1981).

Heritage discourses

Interpretive signage within Red Hands Cave (Figure 3.10) claims that it "has great significance to *all Australians* today … [and is] therefore an irreplaceable part of *Australia's cultural heritage*" (emphasis added). This is a text-book case of what Denis Byrne (1996) describes as a type of heritage discourse that seeks to construct a shared Australian national identity by claiming the material cultural heritage of Aboriginal peoples as the heritage of the nation, thereby ascribing non-Indigenous Australians a "deep history". However, by claiming the cave as the heritage of "all Australians", the

Figure 3.10 "Red Hands Cave Today" Interpretative Sign (located within Red Hands Cave). Photograph by the author.

discourse of heritage becomes a postcolonising force, diminishing the claims of the Dharug and Gundungurra peoples to this place. Heritage conservation practices such as installing a metal grate around the cave with slots for viewing and taking photographs (Figure 3.11), and placing seating opposite it, suggest that this place is being presented as a museum-like object for viewing, not as one with continuing cultural and spiritual significance to Aboriginal peoples.[6]

This framing emphasises the strong settler-colonial narrative within the interpretive signage associated with the cave, a narrative that portrays this place as a "relic" of a disappeared people who no longer occupy this landscape or have ties to this place. Furthermore, the regulatory sign within the cave defines the hand prints and other marks as objects of study, their purpose being to assist "*us* to understand" this supposedly past way of life (see Figure 3.9, emphasis added). The rationale for protecting this place is not based on its value to contemporary Aboriginal peoples, who may be the descendants of those who made the hand prints, but Western science and miscellaneous white others, with no direct connection to the place.

Heritage at risk?

The regulatory sign within the cave (Figure 3.9) encourages visitors to show "*care and respect* for this place", undermining the claim that the cave is of "*great significance* to all Australians", as no one would damage something that was of value to them. This points to the fact that settler-colonial discourses are not necessarily characterised by logic or internal coherence

Figure 3.11 View of Red Hands Cave showing metal enclosure, BMNP. Photograph by the author.

(Blu Barnd 2017). This phrase, in fact, suggests the opposite, that some visitors are likely to disrespect this place, and this is reinforced by the metal grate around the cave, which can be read as providing protection from an active threat. The use of the word "protected" in signage within the cave that states, "*[p]rotected* within the cave is a group of Red Hand prints and stencils" (emphasis added), begs the question, protected from whom or what? The prohibition in the regulatory sign against damaging the cave, reinforced by the penalty for doing so, heightens the perception that this place is under threat of some kind. This discourse of threat and danger suggests a hidden conflict associated with this place, and that this is not made explicit, points to the discursive limitations of this text.

The unspoken subtext is that this risk of disrespect can only be linked to vandalism arising from white racism, which would explain an NSW NPWS employee's interview reference to this place as a "sacrificial site" because it is promoted and accessible to the public, and is therefore susceptible to damage by this same public. The racist graffiti scraped into the metal frame of Figure 3.6, which reads "fuck Abos (sic)" and "I'm not sorry", suggests that racism remains a strong force in contemporary Australian society (Hage 2000). This issue was taken up by three young Aboriginal women I interviewed at the cave, one of whom explained that she approved of the metal grate around the cave because of the graffiti damage she had witnessed at other Aboriginal rock art sites (Chapter 6). The graffiti defacing the frame of

Figure 3.6 and the deteriorated condition of the signage installed on the walk to the cave, taken together, communicate a level of disregard for the cultural heritage of the Aboriginal peoples of the Blue Mountains, both on the part of the public and the NSW NPWS as the managing authority.

Prohibited discourse

There is a striking disjuncture between the narrative of Dharug life told in the Red Hands Cave signage and the current status of this place as part of a national park, accessed by non-Aboriginal people for recreation purposes. However, there is no explanation of what has intervened to cause this change, leading to the supposed absence of Dharug people from this landscape. That the white invasion of Australia is a prohibited or taboo discourse in this signage is apparent in the absence of any discussion of the settler-colonial history of the Blue Mountains, including the impact of this incursion on the Aboriginal peoples of the region. The impact of colonisation is euphemistically alluded to in Figure 3.10, located within the cave, in the claim that the "original meaning of the hands has been *lost*" (emphasis added). If their meaning has, in fact, been lost, which was disputed by the young Aboriginal women I interviewed at the cave, it is not because it was carelessly misplaced, but because of the profound dislocation and ongoing cultural genocide caused by white invasion. The use of the term "Europeans *settled*", in the sentence "most, if not all [of the hands], were placed here before *Europeans settled* in the Blue Mountains" (emphasis added, Figure 3.10), can also be read as a euphemism for invasion and land theft, a prohibited discourse this signage is reluctant to broach.

Jenolan Caves based interpretation

In this section, I provide a final discourse analysis focusing on two audio tours on a free mobile phone application (app) released by the Jenolan Caves Reserve Trust (the Trust) in 2010.[7] The subject of the two tours, the Aboriginal Culture and History Tour and the Healing Waters tour, is the culture and history of the Gundungurra Traditional Custodians of the Jenolan Caves. These tours are the primary form of interpretation of the Aboriginal cultural heritage of the Jenolan Caves available to visitors, although several murals and some limited and difficult to access on-site signage also focus on this theme. Staff of the Trust reported that Gundungurra Traditional Custodians were consulted in the development of the Healing Waters tour, and took the lead in developing the Aboriginal Culture and History Tour. Gundungurra Elders I interviewed recalled approving these tours for use and knew the two Gundungurra people who speak in the audio, one of whom also produced the artwork for both tours.

These tours have both an auditory and a written component (able to be viewed on the app), which are identical. As the content of both tours is too long to provide in full, text fragments considered representative of the whole based on the author's subjective determination after reading the full text have been selected for detailed analysis. Images that accompany the text are described rather than shown because they were produced by a Gundungurra artist and are subject to copyright.

Discourse position

There is a single discourse position that emerges in the Healing Waters audio tour: that of the Gundungurra Traditional Custodian claiming an ongoing and unceded relationship with the Jenolan Caves as Gundungurra Country. As a voice from the margins that in this tour has moved to centre stage, Jäger and Maier's (2016) argument that a singular discourse position suggests a hegemonic discourse, as was the case for Red Hands Cave, does not apply in this instance, suggesting that the operations of discourse are necessarily context-specific. In this tour, Gundungurra speak together as a group, although the statement that "flowing water represents women" and the fact that a Gundungurra woman narrates the tour, places an emphasis on the female voice.

The Traditional Custodian authorial voice is apparent in the Healing Waters audio tour in the repeated use of the plural possessive pronoun "our" and the plural first-person pronoun "we" to refer to the speaking subject of the text. For example, under the component entitled "Trees", the narrator explains that "Numbuk yabbun or smoking ceremonies are very important to *our* culture. When *we* or guests *enter or leave* country *we* hold a numbuk yabbun" (emphasis added). The way in which the Healing Waters tour represents a distinct Gundungurra perspective challenges the assumption of commentators like Tilden (2007) that protected areas should present a universalistic message, hinging around environmental conservation. Although both Indigenous tours on the app do impart environmental messages, this is done from a distinct Gundungurra perspective, and, as discussed later, includes a historical account of the colonisation of this region that is also told from a Gundungurra positionality.

Such content steps outside of the traditional boundaries of protected area interpretation, foregrounding marginalised perspectives, and avoiding the type of simplistic messaging favoured by the American National Park Service Advisory Board (ANPSAB), discussed in the introductory literature review. Like the A<u>n</u>angu authorial voice in the interpretive signage at Ulu<u>r</u>u-Kata Tju<u>t</u>a National Park, the tone taken throughout the tour is instructional, that of a teacher imparting knowledge of language, Country and culture to a student. It seems clear then, that both A<u>n</u>angu and Gundungurra Traditional Custodians, like Tilden (2007), view the purpose of protected area interpretation, at least in part, as educational. However, rather than attempting to reveal so-called universal truths, they speak from a perspective of embodied relationality with kin and Country.

Although there are two distinct narrators within the Aboriginal Culture and History Tour, a singular discourse position is expressed because the same ideological perspective is conveyed by both. The first authorial voice is that of the Gundungurra Traditional Custodian who claims the Jenolan Caves Reserve as their traditional lands, while the second is a non-Aboriginal narrator, whose voice resembles that of a news reader. His status as a non-Aboriginal person is clear from his references to Aboriginal peoples using the third-person plural pronouns "they" and "their". This second narrator, who one visitor interviewee humorously referred to as "Siri", narrates more controversial segments of the tour such as the section dealing with Aboriginal "durella" or resistance to white invasion, and the Europeanisation of Australian environments by settlers. Like the visitors interviewed about the tour, who found this narrator jarring, I too found the use of this narrator an odd choice in a tour designed by Gundungurra Traditional Custodians. However, it is possible that a non-Aboriginal narrator was chosen for the more confronting segments of the tour to attempt to frame these messages as normative and make them more palatable to a non-Aboriginal audience.

Discourse strands

Sovereignty

The primary discourse strand of the Healing Waters audio tour is that Gundungurra Traditional Custodians are the people who belong to the Jenolan Caves and that, despite colonisation, maintain a sovereign relationship with this place. Rather than claiming to own this Country, in accordance with a Eurocentric ontology, they establish their claim to belong to it through their intimate knowledge of it. The discourse of the tour thus successfully conveys a fluid but ongoing Aboriginal onto-epistemology intimately tied to Country.

The Healing Waters tour commences with the voice of a female Gundungurra Custodian speaking a welcome in the Gundungurra language, which is then repeated in English. The narrator states "We would like to welcome you to this sacred land and acknowledge and pay our respects to the Burra people and the Gundungurra Nation. We say 'Welcome and Thank You'". This greeting resembles a "Welcome to Country" offered by Traditional Custodians at the commencement of important meetings and events, based on the traditional Aboriginal practice of seeking permission to enter the Country of another Aboriginal Nation. The important relationship that the Gundungurra have with the caves is indicated by the act of providing a welcome, as it is only those who have a claim to a place who have a right to welcome visitors to it. This is reiterated in the final words of the tour when the narrator thanks visitors "for coming on this journey and sharing some of our culture, *on this our land-past-present-and future*" (emphasis added). The narrator unequivocally claims that the "traditional owners of these caves are the Gundungurra

Nation, *on whose land we stand*" (emphasis added). The discourse position taken is not that of a neutral or impartial commentator, but that of an authoritative Gundungurra custodian asserting that the Jenolan Caves are unceded Gundungurra land.

Although a welcome to Country is also provided by A<u>n</u>angu in the Palya sign at Ulu<u>r</u>u (Figure 3.1), the Healing Waters tour places significantly more emphasis on asserting the Gundungurra's sovereign relationship with the Jenolan Caves. This may be a response to the failure of settler-colonial Australia to acknowledge the ongoing relationships of Aboriginal peoples to the heavily colonised landscapes of south-eastern Australia (Fredericks 2013), unlike Ulu<u>r</u>u-Kata Tju<u>t</u>a National Park, to which A<u>n</u>angu hold title. The theme of sovereignty is also strongly asserted in the Aboriginal Culture and History tour, not only in the welcome and conclusion but in the account of widespread Aboriginal resistance to white invasion from its earliest years.

Cultural continuity

The welcome to the Aboriginal Culture and History tour similarly emphasises not only the longevity of Gundungurra culture, a narrative associated with authenticity discourses, but, like the Healing Waters tour, foregrounds its continuity, explaining,

> On our journey today ... you will hear about the special place that the Jenolan Caves have long been for the local Aboriginal People and learn something of their culture, which *stretches back through tens of thousands of years of tradition,* but which is still *very much a living culture today*. (Emphasis added)

The author of this text is thus aware of and is responding to the myth that the culture of Aboriginal peoples of this region is no longer animate. There is more emphasis placed on the idea of cultural continuity in this text than in interpretive signage within Ulu<u>r</u>u-Kata Tju<u>t</u>a National Park, where the idea of continuity was generally implicit in the discussion of ongoing cultural practices. These differences may reflect the pressure that settler Australians indirectly place on Aboriginal peoples in more urbanised areas to assert their links with pre-colonial Aboriginal cultures to prove they are "authentically" Aboriginal (Cowlishaw 2010, 2011). The linking of time-depth with continuity is a distinction that sets the discourse of the Aboriginal Culture and History Tour apart from the "Aboriginalism" of Red Hands Cave's interpretation, reflecting both its authorship by the Gundungurra and its more recent development. Narratives of continuity stand in stark contrast to those that imply the extinguishment of the relationships of contemporary Aboriginal peoples with protected area landscapes, apparent in interpretation that is not authored by Indigenous peoples (Ballantyne 1995; Carr 2004; Huggins et al. 1995; Staiff et al. 2002).

The narration of the Healing Waters tour by a young Gundungurra custodian emphasises the living nature of Gundungurra culture, offering a sense of contemporaneity. The use of present tense in the explanation of *Numbuk yabbun* in the Healing Waters tour, serves to further emphasise the continuity of traditional practices such as the smoking ceremony. Additionally, the use of the infinitive verbs "enter or leave" in the second sentence of the *Numbuk yabbun* excerpt, indicate that this action has no time limitation, and thus continues to occur in the present and future. Linguistic devices such as these appear throughout the Healing Waters tour, and whether or not used intentionally, have the effect of emphasising that Gundungurra culture is ongoing. Themes of cultural continuity and revival often figure prominently in Indigenous cultural tourism products designed by Indigenous peoples, acting to counter the settler-colonial discourses of erasure, evident in interpretive signage authored by non-Aboriginal people, including at Red Hands Cave (Bunten and Graburn 2018).

Gunyunggakinlung (the Dreaming) and an Aboriginal worldview

The Healing Waters tour also places considerable emphasis on the importance of Gunyunggakinlung or the Dreaming[8] in understanding the contemporary landscape of the Jenolan Caves. The narrator authoritatively describes the actions of creator beings as literally creating the landscape in ways that can still be observed, similar to accounts of Tjukurpa provided in interpretation within Uluṟu-Kata Tjuṯa National Park. For example, the Grand Arch and the Blue Lake segment of the audio tour recount that the arch "was made in the Dreamtime (sic) by Gurungatch … who in a fight with Mirrigan … dug his way through the mountain. At this spot he rested … the blood from his wounds you can see still staining on the rocks above". This text foregrounds an Aboriginal ontology, which is presented as literal fact, and is at odds with a modernist consignment of Aboriginal creation accounts to the status of myth. The centrality accorded to a Gundungurra creation account in this text is a decolonial positioning reminiscent of the border thinking advocated by Mignolo (2011) and the revaluing of "colonized or peripheral philosophies" recommended by Dussel (2008, 19–20 in Grosfoguel 2015, 42). Such a positioning can only be achieved in a first-person account such as provided in this audio tour, where the Gundungurra speak for themselves.

The segment "Aboriginal World View" (sic) within the Aboriginal Culture and History Tour proposes a degree of incommensurability between Western ways of thinking and a Gundungurra ontology, "centred around the concept of garang or 'place'- that landscapes … connect the people to the broader spiritual realms … evident in the stories the land had to tell". At the same time, the Jenolan Caves is said to be a landscape where it is possible for all "to appreciate the spiritual connection that such a place can elicit", and the visitor is encouraged to look around and form such a connection with this place. The narrator suggests that "such connections between individuals and the special places of the Earth is (sic) a natural state of existence; one that *we would do well to acknowledge* and encourage in subsequent generations" (emphasis added).

The message of this excerpt is that although an Aboriginal worldview in all of its multi-dimensionality is not accessible to the non-Aboriginal visitor, they can perhaps instead experience a sense of spiritual connection to places such as the Jenolan Caves that exhibit great natural beauty. A limited capacity to experience something of the Aboriginal relationship with the place may thus become available to them. In the final sentence, the connection between humans and the non-human that characterises an Aboriginal worldview is presented as a model for others to follow, repeating the valorisation of Aboriginal lifeways that feature elsewhere in this tour. Consequently, the tour could be taken to endorse a qualified Indigenisation of the non-Aboriginal visitor through cultural tourism, in a similar way to interpretive signage in Uluṟu-Kata Tjuṯa National Park that calls on the visitor to "feel Country". However, there is a distinction between Traditional Custodians inviting visitors to engage with ontologies such as Country and a settler appropriation of Aboriginal ontologies and even identities, as explored in the chapters that follow.

An Aboriginal perspective on post-contact history

The Aboriginal Culture and History tour is unique in the interpretation under discussion in this chapter in offering an assessment of the white invasion and colonisation of Australia from an Aboriginal perspective. Under the heading of "European Arrival", the tour discusses the impact of Europeans crossing the Blue Mountains on the Burra Burra clan of the Gundungurra people and on their neighbours, the Wiradjuri people. It also offers a frank discussion of Aboriginal resistance (durella) to white invasion by the Gundungurra and Wiradjuri peoples, linking this to earlier resistance by the Aboriginal peoples of the Sydney Basin, Western Sydney and the Hawkesbury region. Unlike standard protected area interpretation, the ideological position of this text is overt, rather than masquerading as neutral scientific discourse, in which systems of representation present in the text may be hidden (Staiff 2014). It is also important to recognise that Aboriginal involvement in developing these tours has led to a focus on "truth-telling" about controversial topics such as the frontier wars (Ballantyne et al. 2012).

Presenting Aboriginal resistance as widespread and persistent provides a counter-discourse to the discredited colonial myth of "terra nullius", that Australia was an empty land peacefully settled by the British (Banner 2005). The Aboriginal leader, Pemulwuy, is described in positive terms as a "great resistance fighter", the use of this term evoking the concept of a just war against oppressors. Twentieth-century resistance movements were primarily uprisings by peoples of the colonised global south against European colonial powers, and the use of the term resistance situates Aboriginal durella as an early example of this. Presenting the colonial government of Australia as a form of illegal "occupation", as in the excerpt below, serves to justify such resistance. This section of the audio tour concludes that "the *fierce Aboriginal resistance* to the *occupation* of their lands, the extinguishment of their languages and the diminishing of their culture is only now being recognised as it *should be*" (emphasis added). This "fierce ... resistance" is expressed as a source of pride, deserving of recognition and respect, contrary to colonial accounts of Aboriginal resistance that characterised Aboriginal peoples as savage, murderous and treacherous (Asten 2013; Dodson 2003).

The use of the modal verb "should be" in this context provides a strong moral direction from the Gundungurra authors of this text that this history of Aboriginal resistance must be acknowledged by contemporary Australians, perhaps calling into question the legitimacy of their territorial claims. This narrative represents a new discourse strand that the white "occupation" of Australia is unjust and was actively resisted by Aboriginal peoples from the early decades of the colony. It forms a counter-narrative to arguments that the colonisation of Australia was justified on the basis that the country was "terra nullius" or empty land. In so doing, this tour furthers a decolonising agenda by challenging foundational myths about the white settlement of Australia.

Conclusion

The analysis provided in this chapter suggests that the degree of involvement and control that Aboriginal peoples have over the interpretation of their cultural heritage in protected areas has a profound impact on the messages conveyed. On-site signage within Uluṟu-Kata Tjuṯa National Park and the Aboriginal themed audio tours on the Jenolan Caves app, developed in consultation with or by Traditional Custodians, feature a Traditional Custodian voice and discourse position that contrasts with more dated interpretation such as at Red Hands Cave, which excludes it. Interpretation over which Aboriginal people have been able to exercise control also represents traditional aspects of culture as continuing, just as the relationship with Country is said to be ongoing. This is in stark contrast to the on-site signage at Red Hands Cave, which positions an Aboriginal relationship with culture and Country as of the past.

Furthermore, a politicised Aboriginal voice is clearly apparent in the Jenolan Caves' Aboriginal Culture and History tour in the discussion of Aboriginal resistance to white incursion and settlement. This voice transgresses the prohibition against breaking "the great Australian silence" (Stanner 2009, 189) about white invasion and frontier violence that is evident in state-sanctioned interpretation in both Uluṟu-Kata Tjuṯa and the Blue Mountains National Parks (Red Hands Cave). As a result, it is evident that discursive limitations are not merely a product of the era in which interpretation was developed but also result from who is responsible for authoring it and the level of control that Traditional Custodians have over interpretive content.

Despite changes in these interpretive discourses over time, reflecting the positive self-assertions of Aboriginal peoples and changing understandings of post-contact histories, there are certain persistent discursive features across government-sponsored interpretation that remain postcolonising. Appearing in protected areas with World Heritage listings, these include heritage and natural resource management discourses that continue to separate natural from cultural heritage and those that rely on a narrative of authenticity to promote the time-depth of Aboriginal cultures. Relatedly, scientific and natural resource management discourses define Country as a resource to be managed and impose Eurocentric regimes of classification upon the non-human others who share Country (Howitt et al. 2013; Searle and Muller 2019). These discourses, taken together, have the effect of distancing the research locales from Aboriginal onto-epistemologies that are otherwise emphasised through the primacy accorded to the concepts of Tjukurpa and Gunyunggakinlung. In so doing, these Eurocentric discourses undermine the decolonising orientation of interpretation within Uluṟu-Kata Tjuṯa National Park and on the Jenolan Caves app.

However, Traditional Custodians sharing their belief systems and knowledge of Country with non-Aboriginal visitors can also be an empowering act, one in which the traditional power hierarchy between coloniser and colonised is reversed, as Traditional Custodians possess valued knowledge that visitors lack and wish to access (Travesi 2018). This suggests that, as Bhabha (2004) has argued, the roles of colonised and coloniser are not fixed and unchanging, even in a landscape such as the Jenolan Caves that has been significantly altered by settler-colonial incursion. Provision of cultural interpretation that is fully authored and controlled by Traditional Custodians, because it reverses the power relationship between colonised and coloniser, can thus function as a form of decolonisation. As discussed in Chapter 4, there is the greatest potential for this in Traditional Custodian led cultural tours than in fixed locale-based signage because Traditional Custodians have full control to direct the tour as they see fit, utilising the Aboriginal cultural practice of person-to-person oral knowledge transmission. These and other themes are taken up in the chapters that follow, including in Chapter 4, which explores the goals of Traditional Custodians for Aboriginal led cultural experiences and provides an account of my participation in tours and other cultural activities within a number of research locales.

Notes

1 For example, the Australia ICOMOS Statement on Indigenous Cultural Heritage, 2001; the Australia ICOMOS Practice Note – Interpretation, 2013; the NSW Heritage Office's Interpreting Heritage Places and Items, Guidelines, 2005; the Australian National Trust's 'We're a Dreaming Country' – Guidelines for Interpretation of Aboriginal Heritage 2012 and the ICOMOS Charter for the Interpretation and Presentation of Cultural Heritage Sites, 2008. The United Nations Declaration on the Rights of Indigenous Peoples, 2007 gives specific attention to the rights of Indigenous people to practice, control and pass on their cultural traditions (see in particular Articles 11, 12 & 13).

2 The opposition of the NT Government to the return of the park to Aṉangu, also known as "hand-back", was briefly covered in a video playing in the Uluṟu Cultural Centre during my 2019 fieldwork, that Parks Australia refers to as the "hand-back" video.

3 More recently developed interpretation close to the Glenbrook entrance of the park is very different from the interpretation discussed above in that it acknowledges that Blue Mountains Traditional Custodians continue to regard places such as Red Hands Cave as of cultural importance and provides a welcome to the Park from them phrased in the present tense, although not in the first person. Unfortunately, the out of the way location of this signage on the wall of the Glenbrook NSW NPWS office, which is closed to the public, means that most visitors to the cave are unlikely to see it.

4 Daruk is a variation on the spelling of Dharug. I use the spelling "Dharug" based on the spelling that appears in the AIATSIS map of Indigenous Australia, https://aiatsis.gov.au/explore/map-indigenous-australia.

5 Merriam-Webster online dictionary, https://www.merriam-webster.com/dictionary/relic, accessed 13/01/2022.

6 I am indebted to Dr Steve Brown for this insight.

7 The app was produced by Acoustiguide Interactive, a company that specialises in developing interpretation accessed electronically through devices such as mobile phone and tablet applications, interactive kiosks with touchscreens and immersive websites, not only in open spaces but in museums and galleries. There are six tours in total on the app, but only two focus on the Gundungurra heritage of the caves. The app that includes these tours can be downloaded at https://appadvice.com/app/jenolan-caves/894486581.

8 The idea of the Dreaming is similar to the Aṉangu concept of Tjukurpa. The term the Dreamtime was a translation of Tjukurpa and related terms in Western Desert dialects by early anthropologists such as Baldwin Spencer and Frank Gillen "to denote the mythic primordial times of Aboriginal religion" (James 2015, 36) and was taken up and has been used widely in Australia since 1899, including by some Aboriginal peoples. The term the Dreaming is now preferred as it is more suggestive of the ongoing nature of creation events in Aboriginal spirituality.

4 Interruptions/Revivals/Reinscriptions

Traditional Custodian led cultural tours

Introduction

Indigenous tourism has been defined "as tourism activity in which indigenous (sic) people are directly involved either through control and/or by having their culture serve as the essence of the attraction" (Ruhanen et al. 2015, 74). In this chapter, I discuss four guided experiences led by Aboriginal peoples, focusing on tours that are not just about Aboriginal peoples, but authored by them, as this distinction is crucial to my research. Two of these tours took place in or near Uluṟu-Kata Tjuṯa National Park (Patji and Maraku Arts cultural tours), and two occurred in NSW, one at Yellomundee Regional Park (Yellomundee Culture Camp) and the other on Crown Land in the lower Blue Mountains (Blue Mountains Walkabout).

These tours were included in the research to determine their relative impact on visitors vis-à-vis the interpretive signage and other forms of interpretation discussed in Chapter 3. In so doing I hoped to determine whether Traditional Custodian led tours allow for a deeper and more meaningful engagement with Aboriginal cultural heritage in protected areas than non-interactive forms of interpretation such as on-site signage. In the process of conducting the research, my interest in these tours further evolved to consider questions of Indigenous agency in cultural tourism and the implications of this for the broader question of decolonising tourism in protected areas. Relatedly, I investigate whether Aboriginal led cultural tours aim to open up Aboriginal ways of being to non-Indigenous participants, including ontologies of Country and relationality. I also consider whether any limitations are placed on access to these ontologies by tour providers, critically assessing the meaning of processes of Indigenisation in a settler-colonial context (Veracini 2015). Additionally, I assess whether the tours on offer, in both conception and delivery, *interrupt* normative narratives of Aboriginality arising in a settler-colonial context, such as discourses of authenticity, or *reinscribe* them.

In order to explore such questions, I outline the Aboriginal cultural tour operators' goals for their tours based on my interviews with them and their online and printed promotional material. The views of non-Aboriginal people who in some cases partnered with Aboriginal providers to offer these

DOI: 10.4324/9781032654973-4

tours, such as interpreters and drivers/guides, are also included. Introducing the perspectives of Aboriginal tour providers on the experiences they offer visitors not only addresses a gap in the tourism scholarship (Carr et al. 2016) but also centres Aboriginal perspectives as important, to some extent balancing my focus on the narratives of mainly non-Aboriginal visitors in the remaining chapters of this book.

The bulk of this chapter is written in a more personal and descriptive style than previous chapters and those that follow, because the primary data for this chapter are my own field-work notes describing the cultural tours I participated in and my observations of the participation of other visitors. These observations are signalled by italicised body text, which is not only descriptive, but also provides an insight into my critical affective responses to these experiences, enabling a direct exploration of my own positionality in relation to the research.

Indigenous led cultural tours as a form of interpretation

As this chapter discusses cultural tours led by Aboriginal peoples as a form of interpretation, it is important to first briefly survey the scholarship on guided interpretation per se and Indigenous led tours in particular and also to revisit some of the related conceptual issues. A key focus of the interpretation literature is the relative effectiveness of different interpretation methods in terms of length of time visitors devote to reading or listening to content, interactive versus passive methods, "hot" versus "cool" interpretation (Uzzell 1989), and knowledge transmission and retention. Compared with static, passive forms of interpretation such as signage, guided interpretation is often considered better at engaging the visitor as it has the potential to foster an intense and interactive experience, combining the "active delivery of messages with a multi-sensory, complex social interaction" (Hughes and Morrison-Saunders 2005, 161). Hughes and Morrison-Saunders (2005, 161) speculate that the use of guides "may act to draw visitors in and shape messages to suit personal contexts". Furthermore, Wearing and Neil (1999) note that interpersonal communication can respond to changing contexts, diverse audience needs and spontaneous events, allowing a dynamic two-way interaction that can potentially exert more influence on the visitor. It is also argued that guided interpretation that involves physically moving through the landscape lends itself to embodied experiences that visitors find more meaningful, memorable and transformative (Beck and Somerville 2002; Wright et al. 2012).

While guided interpretation of Aboriginal cultures can be provided by both Indigenous and non-Indigenous guides, Traditional Custodians are the authority on their own cultures, and thus typically in a better position to communicate cultural knowledge to visitors than non-Indigenous people. Traditional Custodian led cultural tours also have a number of advantages from a decolonial perspective. Tours that are provided by Aboriginal and Torres Strait Islander owned enterprises can have economic benefits to communities, although such benefits have not proved to be as significant as Australian governments (and Indigenous tour operators) have hoped (Ruhanen et al. 2015). Perhaps more significantly, these tours reverse settler-colonial power relations by encouraging visitors to recognise Aboriginal and Torres Strait Islander authority and knowledge of Country (Travesi 2018). This places the Indigenous tour provider in an empowered position in relation to the visitor and by promoting knowledge systems labelled marginal, has the potential to challenge postcolonising systems of power (Mignolo 2011; Quijano 2007). I noticed a similar power reversal in the way that Anangu labelled visitors who climbed Uluru *minga*, a Pitjantjatjara word that means "ants". This epithet was not only a visually accurate description of the appearance of climbers, spread out in a black line across the spine of Uluru, but an act of imaginative reframing, that perhaps minimised their power to claim attention, in response to the cultural offence they caused by climbing Uluru.

There is also considerable evidence that Indigenous tourism can play a role in what Ruhanen and Whitford (2019, 180) describe as the "maintenance, rejuvenation and/or preservation of Indigenous cultures, knowledges and traditions" and "can help sustain intangible elements of Indigenous cultural heritage, including languages, stories, song, art, dance, hunting methods, rituals and customs". In the settler-colonial context, Indigenous cultural tourism can thus be regarded as part of what Denis Byrne (2003a, 77) describes as a process of return to Country that involves not only the repatriation of Aboriginal human remains but also a "return of Aboriginal visibility in the colonised landscape". In Uluru-Kata Tjuta National Park, Anangu's ongoing relationship with Country is made visible not only through Anangu-led cultural tours but also through paintings, baskets dyed and woven from materials sourced from Country and wood carvings, which are sold to visitors via Anangu owned business in the Uluru Cultural Centre, representing both an income source and a way to continue practicing culture.

In the international context, a number of authors discuss the improved status of Indigenous peoples such as the Emberá of Panamá and Columbia (Theodossopoulos 2018), the Pewenche of Chile (Palomino-Schalscha 2018) and the San of Botswana (Giraudo 2018) as an indirect consequence of their involvement in cultural tourism. For example, Theodossopoulos (2018) discusses the revitalisation of the dances of the Emberá of Panama as a consequence of tourism, and how this also led to the revitalisation of traditional dress among women and changing attitudes to this attire by their Latino neighbours. This valorisation of Indigenous culture has occurred not just at the broader societal level but also in the tourist encounter, leading to a renewed cultural pride among Indigenous communities (Stocker 2018). Indigenous cultural tourism thus enables Indigenous lifeways and cultures to be depicted in a positive light, reversing the pervasive deficit discourse on Aboriginality identified by Fforde et al. (2013), discussed in the following chapter.

A further consequence of Indigenous peoples' involvement in Indigenous tourism is that the act of representing an Indigenous identity can lead them to renegotiate the ways they think about themselves, sometimes in relation to other Indigenous peoples. This can contribute to forms of politicisation, prompting Indigenous peoples to claim a pan-Indigenous identity to support a discourses of rights, particularly around claims to traditional lands and other resources such as water. Palomino-Schalscha (2018) explains how the Pewenche do not shy away from these topics in their interactions with visitors. Indigenous led cultural tourism thus gives Traditional Custodians control of the type of interpretation provided and can mean that subjects excluded from state-sponsored narratives, such as the continuing impact of settler-colonial interventions on Indigenous peoples, are broached.

Some Indigenous tourism experiences also frame interpretive content within Aboriginal onto-epistemologies that emphasise relationality with the more-than-human and the agentic nature of Country, rather than neoliberal Eurocentric scientific discourses of natural resource management and heritage narratives of natural and cultural values (Bawaka Country et al. 2013; Palomino-Schalscha 2018). Valorising and prioritising Aṉangu onto-epistemologies within cultural tourism, as I argue occurred on the Patji tour, discussed below, can represent the type of epistemic decolonisation called for by decolonial scholars such as Quijano (2007) and Mignolo (2011) (see Chapter 2). However, the literature also raises concerns that Indigenous tourism can lead to the commodification of Indigenous cultures and the reification of stereotypes about Indigenous peoples (Bunten 2018; Carr et al. 2016). For example, Kanak scholar Haunani-Kay Trask (1991, 161) is highly critical of (mis)representations of Hawaiian culture by the tourism industry as "hula dances, ukeleles and pineapples", in the context of this industry's neo-colonial presence in Hawai'i (see also Trask 1999). Similarly, Aluet scholar Alexis Bunten (2018, 11) cautions that the "particular imaginary [that] draws visitors to the tourism site ... can also trap Indigenous hosts into representing themselves in ways that do not resonate with the complexity of their contemporary lives".

But questions of identity and self-representation are complex matters. Postcolonial scholar Gayatri Spivak is well-known for arguing, against the post-structuralist canon, that when it comes to the majority world woman reclaiming a marginalised identity, "the risk of essence" (1990, 126–7) may need to be taken, however, the decisive factor in the legitimacy of this enterprise is "who it is that uses the strategy". The positive reframing of essentialist stereotypes by Indigenous peoples is thus fundamentally different from the application of such epithets by the non-Indigenous. Going further, Moreton-Robinson (2015, 12–13) critiques postcolonialism's rejection of essentialism (except when deployed in strategic ways), arguing in favour of Indigenous ways of knowing based on an ontological relationship with the land and declaring claims of essentialism the imposition of a universalistic Western worldview.

A nuanced understanding of Indigenous self-representation in the context of Indigenous cultural tourism is thus required, including an appreciation that essentialist self-representations may enact individual and community ascribed meanings. For example, Stocker (2018, 86) discusses a dance performed by a young man who is part of the Chorotega people of Costa Rica at a cultural festival, in which he wore body paint "in the style of Hollywood Western (mis) respresentations of US Plains Indians". As Stocker (2018, 86) notes "[w]hile the dancer, was, in fact, performing stereotype, it was an image that may have become meaningful to some locals through its ability to index and embody expectations of indigeneity (sic)" (even if not their lived reality). Stocker concludes that such stereotypes can be turned to the benefit of this community, become meaningful to them and in a sense undergo forms of transformation. Such a reading aligns with Clifford's (2013, 61–62) understanding of Indigeneity as a process of articulation, in which "cultural forms will always be made, unmade and remade". In this context, the question of whether stereotypes are problematic, just like the question of whether cultural practices are authentic or inauthentic, is best self-defined by Indigenous peoples.

The literature surveyed above thus cautions about or rejects essentialist stereotypes when they are externally imposed by powerful, sometimes colonial forces, including to meet the perceived expectations of the tourist industry. However, it also recognises the strategic value of the integration of essentialist stereotypes into Indigenous identities in certain circumstances and goes so far as to defend these when they express core Indigenous ontologies, such as an inherent relationship to Country. I agree with this nuanced understanding of what has been termed "essentialism" in poststructuralist theory as it applies to contemporary negotiations of Indigeneity and seek to highlight how these different expressions manifest in the cultural tours described in this chapter. While some Traditional Custodian led cultural tours may be informed by settler-colonial discourses, such as discursive formations that link Indigenous identity with a timeless authenticity, other tours position their narratives within a settler-colonial historical context and discuss the contemporary realities of Aboriginal peoples.

For example, cultural experiences such as Patji, discussed below, emphasised both cultural continuity in the maintenance of traditional knowledge of bush foods and the medicinal uses of plants but also the way that A<u>n</u>angu have been impacted by a history of colonialism, up to and including key twentieth-century events. Other cultural experiences discussed below, such as the Yellomundee Cultural Camp, enacted a revival of traditional cultural practices including the smoking ceremony, dance, ochreing the body and actively caring for Country through weed removal and cultural burning. The Traditional Custodians who led the camp also modelled an Aboriginal ontology of relationality and a belief in the agentic nature of Country (Dudgeon and Bray 2019; Tynan 2021). However, non-Aboriginal participants were invited to share in these practices, or in other words, to Indigenise themselves, in ways that may have blurred the boundaries between an Aboriginal sovereign relationship with Country and the type of relationship available to settler subjects. The remainder of this chapter describes each of these cultural experiences from my perspective as a participant observer, closing with a discussion that addresses the complex issues of Indigenisation and cultural appropriation, Aboriginal sovereignty and authenticity debates within the context of Aboriginal led cultural tours.

Patji tour, A<u>n</u>angu, Pitjantjatjara, Yankunytjatjara Lands, 2019

Patji is the name of the A<u>n</u>angu, Pitjantjatjara, Yankunytjatjara (APY) lands south of Ulu<u>r</u>u-Kata Tju<u>t</u>a National Park where the SEIT Outback Australia tour took place, although visitors were not allowed to know what the word means. This was one way in which A<u>n</u>angu onto-epistemologies structured the tour offering, reflecting the importance of restrictions on cultural knowledge within A<u>n</u>angu society. The Patji tour was a collaborative offering by SEIT, a non-A<u>n</u>angu small group tour company based in Yulara, in partnership with the A<u>n</u>angu business, Uluru Family Tours. The Ulu<u>r</u>u family is an important family of Traditional Custodians with responsibilities towards Ulu<u>r</u>u-Kata Tju<u>t</u>a National Park and surrounding lands. The tour group I accompanied comprised six older Anglo-Australians, two of whom agreed to participate in an interview about the tour at their hotel the following day. I also interviewed the A<u>n</u>angu host and the tour driver/guide employed by the tour company.

The Patji tour was marketed in a promotional flyer as a "*True* Aboriginal Experience", promising visitors an opportunity "to learn about the *oldest living culture in the world* … [and to] hear stories such as how Paddy Ulu<u>r</u>u fought for Aboriginal land rights to where Ulu<u>r</u>u is today" (emphasis added). It promised that "the Ulu<u>r</u>u family will share their personal story with you during this *authentic cultural experience*" (emphasis added). The tour was thus promoted on the basis of a discourse of authenticity, signalled not only by the direct use of this term but by reference to the time-depth of A<u>n</u>angu culture and the personal nature of the encounter offered with a member of the Ulu<u>r</u>u family. It seems likely that these themes were promoted as they are familiar tropes of Aboriginality within a tourist context (Beck and Somerville 2002; Seiver and Matthews 2016). The name Ulu<u>r</u>u would be a very familiar one to visitors, as it is shared not only by the national park but its most famous feature, the large red sandstone monolith that rises dramatically from the flat desert plain.

When I asked the A<u>n</u>angu host of the tour, a softly spoken young man of few words, what the purpose of the tour was from his point of view, he offered:

> Show them bush tucker; what's out there. Also, important to tell them the history of the [Ulu<u>r</u>u] family. They may have a little bit of respect for culture, ceremony; warn them not to take photo of sacred site, not to walk in sacred area. Let them know what's out there … See the backyard. See the Country.

From the perspective of the more voluble driver/guide provided by SEIT, the purpose of the tour was:

> … really about learning from the Uluṟu family, giving guests the experience of being with the family that is tasked through culture with looking after this place and passing down their stories … whilst the tour that we do doesn't go into a lot of Tjukurpa … there's also what the Aṉangu refer to as family Tjukurpa, which is the story of Paddy [Uluṟu], giving guests the opportunity to meet the people behind the rock,[1] the people whose ancestors have been out here for so many generations and are still here, giving them an understanding of … not just the rock, the whole landscape.

These explanations make it clear that the tour was not simply intended to be an opportunity for visitors to see the sights, but for them to come to a greater depth of understanding about Aṉangu culture and the meaning of the entire landscape, at least those aspects that could be shared. It was also intended to give visitors an insight into the history of the Uluṟu family and to engender change in visitors by encouraging greater respect for Aṉangu and their culture. These themes of transformation echoed the purpose of other Indigenous led cultural tours, such as those offered by Bawaka Cultural Enterprises in Australia's NT (Bawaka Country et al. 2017). Like the interpretive signage in Uluṟu-Kata Tjuṯa National Park discussed in Chapter 3, the Patji tour conveyed to visitors that when they were on Aṉangu land, Traditional Custodians such as the Uluṟu family expected them to behave according to Tjukurpa by not photographing sacred sites or walking in sacred areas. The Patji tour thus prioritised the primacy of the Aṉangu onto-epistemology of Tjukurpa, including communicating that visitors were encompassed by this ontology when they were on Aṉangu land. This tour therefore also conveyed an ongoing Aṉangu sovereignty over Country and the knowledge and authority of Aṉangu in their relationship with the APY lands.

Field note recollections, Patji tour, May 2019

The morning was fine, but cold for May, when the bus picked me up outside the Uluṟu Cultural Centre and we headed out of the park past the outskirts of the community of Mutitjulu onto a corrugated dirt road. The loud noise of the bus shaking and our bodies shaking along with it during the seven-hour return trip, as my interviewees later commented, was a distinctive feature of the experience. The din of the moving bus kept conversation in the vehicle to a minimum, or perhaps it was my presence as a participant observer.

I found the country outside of the park heading towards South Australia (we were about 40 kilometres from the border), desolate and impossible to love, except in part, when we came across low rocky outcrops, shining in the late afternoon sun (Figure 4.1). The country looked like a fire had gone

Figure 4.1 Rocky outcrops on Patji tour, APY lands, NT. Photograph by the author.

through and left- nothing. It was much more verdant around Uluru. For the first time, I felt some reluctant empathy for "explorers" like William Gosse (1873, 9) who described this land so disparagingly in his journal as "the same wretched country" that he travelled through day after day (Figure 4.2).

While this was my immediate affective response, it was one arising from a settler way of seeing, based on superficial impressions of the landscape. For

Figure 4.2 Panoramic view of Uluru, Patji tour, APY lands, NT. Photograph by the author.

Aṉangu, this Country is "the backyard", a source of food, a place of deep meaning, and one they have caring responsibilities towards, just like the generations of Aṉangu before them. It is also wholly theirs, as it was the subject of a successful land rights claim, and a place tourists cannot enter without a permit.

The Aṉangu host called stops to our jarring progress at various intervals, revealing his intimate knowledge of this land by sharing aspects of Country with us. Communicating his knowledge of Country to visitors was one way for the Aṉangu guide to practice and maintain his cultural knowledge, demonstrating the role that tours such as this have in rejuvenating and preserving Aṉangu culture. The most remarkable encounter for me was with a rock hole that was an important water source for Aṉangu. Our host explained that Aṉangu women had worn down the granite around the hole grinding grass seeds over who knows how many years, a moving "taskscape" (Ingold 1993, 159) of history, etched in stone. The surface was soft to the touch and shiny in parts, catching the glints of the afternoon sun, a quality Belinda Liebelt (2019) associates with grindstones (Figure 4.3). This rock surface was in fact one large grindstone, and as Liebelt (2019, 273) explains, the material properties of such objects/places "reflect the sustained energy and time invested" in them, properties that "cannot be forged or faked". My urge to touch the surface echoed Liebelt's (2019, 275) affective engagement with the grindstones she touched as part of her work as a heritage consultant, offering a sense of connection with the generations of Aṉangu women who used this place.

Another noteworthy stop occurred early in our journey, when our host demonstrated how Aṉangu dig up witchetty grubs. He complained as he dug that this was women's work: "Men get the meat" he said, by way of explanation. This did not imply that Aṉangu women's food gathering roles were less important, far from it, given the ongoing role of traditional foods like witchetty grubs and honey ants as a food source. We passed the witchetty grubs around hand to hand, a close encounter with not only the grubs but our fellow travellers. One of the visitors noticed that the witchetty grubs were cool to the touch, as they'd been in roots of the mulga.

This sense-based experience prompted a vocalised observation on the part of one visitor, while another commented on a very fragrant plant that smelled like lemongrass, passed around the group later in our journey. It reminded her, she said, of the Lemon Myrtle in her Queensland garden. Scent thus has the power to evoke involuntary memory, the type of memory "that is not cognitive but sensorial [as] it emerges in responses to objects" (Witcomb 2013, 268). These sense-based experiences added to the affective dimension of the tour, distinguishing it from museum-based encounters with Indigenous cultures which typically do not permit such intimate engagements with objects in their collections.

At the furthest point of our journey through Patji we arrived at a metal shelter, where our host demonstrated spear throwing and efficiently made a

Figure 4.3 Rockhole on Patji tour, APY lands, NT. Photograph by the author.

camp fire and cooked the witchetty grubs gathered on the way. Since spears are no longer typically used by Anangu to hunt game, the inclusion of spear throwing could be regarded an example of the performance of stereotype (Stocker 2018). However, rather than critically assessing this element of the tour, it is perhaps better seen as the contemporary remaking of an Indigenous cultural practice for the tourist market, as I argue below also occurred in the integration of smoking ceremony into the Yellomundee Culture Camp (Clifford 2013).

The ongoing practice of collecting and eating food from Country, such as witchetty grubs, was also in a sense remade by being incorporated into an Indigenous cultural tourism experience. Being offered food from Country cooked in fire felt like a special type of hospitality. Some of us tried a witchetty grub for the first time, the men more eager than the women, comparing the taste to fried egg. The guests I later interviewed had observed the deft, almost ritualised way our host laid the fire and cooked the grubs, but also commented on his use of matches to light it, as if he had somehow cheated by not lighting it in a traditional way. Rather than recognising the way our host laid and lit the fire as an example of cultural hybridity (Bhabha 2012), these visitors assumed that to be "authentic" a practice should be unchanged, even though it was clear that the method of constructing the fire and cooking the grubs was itself culturally specific.

Later in the tour, while we were sitting comfortably around the camp fire, our host began to tell us about his family Tjukurpa, sharing stories that were both deeply sorrowful and filled with pride. I later realised that telling stories around a camp fire was an Aboriginal way of sharing story, of interpreting culture to visitors, one that I would also encounter at the Yellomundee Culture Camp. Our host took out a folder with A3-size laminated images of family members and began to recount important events in the Uluru family's history, using the pictures to tell a story. He told us about the shooting death of a member of the Uluru family by a police officer at Uluru in the 1930s, about how the exploding of nuclear bombs at Maralinga in the 1950s and 1960s made Anangu sick,[2] *and about the struggle for land rights and 'handback' of Uluru-Kata Tjuta National Park, showing a picture of his ancestor, Paddy Uluru, meeting with politicians*

Our Anangu host also showed a photograph of his grandmother and spoke about the removal of four of her children as part of the government policy of assimilation, now known as the Stolen Generations, and recounted how she walked 200 kilometres to try to get them back. Apart from one of the participants commenting on the length of this journey, guests were generally silent after hearing this story and it was not apparent what they were thinking or feeling. The tour guide later commented that it was not unusual for visitors to receive the host's account in silence, explaining,

> *Australians … [will] often … shut down. Australians are very quiet during that tour in general, they're taking it in, it might be the first time they've heard from a Traditional Owner talking about what happened … All of the guides [hosts] will talk about the Stolen Generations very briefly…. Australians will acknowledge it, probably not in detail, and it will probably be the first time they've been confronted face to face with a Traditional Owner talking about that.*

I was disappointed by the silence and lack of any visible reaction to the Anangu host's account from the tour group and inclined to judge the group for what I interpreted as a lack of empathy. However, when I spoke to the two visitors who agreed to be interviewed the following day, they made it clear that they had a strong response, one that was entangled with emotions such as white shame and guilt, and clear judgements about the immorality and harm of assimilatory policies of Australian governments that led to the removal of Aboriginal children (see Chapter 6*). It thus cannot be assumed that silence equals a flat response. There are different types of silences, those arising from marginalisation and prohibition, and silences that are "performative and productive", for example, "within the processes of remembering" (Mason and Sayner 2019, 17). While it was not possible for me to know the unspoken thoughts and feelings of the visitors who did not agree to participate in an interview, their silence did not necessarily reflect a lack of capacity to be affected by the stories that they heard.*

Maraku Arts tour, Uluṟu-Kata Tjuṯa National Park, 2019

The second cultural tour I participated in with an Aṉangu guide was offered by a medium-sized tour company through an Uluṟu-based, Aṉangu owned, enterprise, Maraku Arts, and took place at the base of Uluṟu. This event formed part of a longer tour of Central Australia that this group was participating in. Unlike Patji, which was a full-day experience outside of the park, this tour lasted approximately one hour and was structured to fit into the tour company's busy schedule. The tour company promoted this tour on its website as an experience with an "Aboriginal and Torres Strait Islander Guide to gain an insight into the land and the local Mala people", one in which they will also "[d]iscover *ancient* cave paintings near Mutitjulu Waterhole as a local Aṉangu guide explains the culture and stories behind the art and its connection to the land" (emphasis added).[3] The familiar trope of the time-depth of Aṉangu culture appeared again in a further discourse plane, that of web-based promotional material developed for tourists, in the reference to "*ancient* cave paintings", signalling the allocation of value to these cultural markings based on their antiquity as part of an authenticity discourse.

The non-Aboriginal guide/driver from the tour company explained that they offered an experience with Aṉangu because:

> ... our company is very ... supportive of the Aṉangu and all Indigenous people of this area. We want to convey and educate the passengers on the influence of the Indigenous people for those areas and how they're taking back control of their country and the conservation efforts in those places.

She also explained that in their interaction with visitors, she had observed that Aṉangu guides "want to teach the passengers that ... [the park] is more than just a rock".

Field note recollections, Maraku Arts tour, May 2019

I met the Aṉangu guide, her interpreter, the touring company guide/driver and the approximately ten guests at the car park for the Mutitjulu waterhole *early on a cold Wednesday morning. We walked together a little way along the path towards the waterhole, and at the request of the A*n*angu guide, who spoke only Pitjantjatjara during the tour, sat down in the dirt in a semi-circle around her.*

*The A*n*angu guide, through her interpreter, taught the guests about some of the bush foods in that area, the former uses of the women's, men's and family caves, and explained the artwork in some of the caves. She also taught guests some of the symbols used in A*n*angu artwork and their meaning, and common animal tracks, like kangaroo, dingo, lizard and snake by drawing in the dirt. Guests I later interviewed recognised drawing in the dirt as a traditional teaching style of A*n*angu, and appreciated being taught in this way. The A*n*angu guide explained the differences between men's and women's tools, such as spears and digging sticks, and passed these around the group. Something about the materiality of handling objects like tools seemed to prompt a deeper level of engagement in the group. A series of seemingly random questions followed about a glue called kiti used in constructing tools, spear throwing and whether A*n*angu could distinguish between different people's footprints. Questions such as this revealed guests' interest in the traditional A*n*angu lifeways that were the focus of the tour.*

*The A*n*angu guide then told one of A*n*angu's important creation stories, the Seven Sister's Tjukurpa, and the interpreter, at the guide's request, provided an account of the Kunya and Liru story, an important creation account associated with the Mutitjulu waterhole, where the tour took place.*

Next, visitors were led by the tour guide/driver on a short walk to the Mutitjulu waterhole and back. Because there was another group ahead of us, we had to form a queue and wait until they left to gain entry. The lack of exclusive access to places like this waterhole highlights one of the disadvantages of mass tourism that visitors later discussed in interviews.

*The final stop on this very structured tour was the Ulu*r*u Cultural Centre, where a non-A*n*angu staff member from Maraku Arts explained more about traditional A*n*angu tools and weapons, and the local practice of wood carving, a continuing cultural practice but one now mainly targeted at the tourist market. Guests were then left to explore the rest of the Cultural Centre, before returning to their bus.*

While the Patji tour connected Tjukurpa stories and food gathering practices with contemporary Anangu culture and post-contact history, and enabled guests to connect with Country in an affective and embodied way, these connections were less apparent on the Maraku Arts tour. This tour focused solely on traditional cultural practices and beliefs, and post-contact historical perspectives were missing. While visitors seemed interested in this content, as Bunten argues in this chapter's literature review (2018, 11), appealing to the tourist "imaginary" of Indigeneity can limit the ways in which Indigenous hosts represent themselves, and such self-representations may not reflect the complexity of their contemporary lifeways. The limitations of this tour may have contributed to an interviewee who participated in the Maraku Arts Tour engaging in a form of authenticity discourse that romanticised remote living Aboriginal people and denigrated those living in urban areas, discussed in Chapter 5 (see Jason's account).

Blue Mountains Walkabout tour, Faulconbridge, 2019

It proved less easy to identify cultural tours suited to my research in the Blue Mountains area than in Uluṟu-Kata Tjuṯa National Park, and not all Traditional Custodians I approached with requests to participate in tours responded. I first learned about the Blue Mountains Walkabout tour through a travel guide to Indigenous Australia published by Yiman and Bidjara scholar Marcia Langton in 2018 and obtained the Aboriginal tour guide's consent and that of his participants to accompany them on a tour in September 2019. The business owner and tour leader described himself as an Aboriginal man living in the Blue Mountains, claiming knowledge about Aboriginal culture and Lore based on initiation by Aboriginal Elders. The website promoting the tour promised visitors a "*genuine* Australian Aboriginal walkabout", offering "*ancient* wisdom to enrich your life and a deep connection to our country & culture" (emphasis added), revealing a similar authenticity discourse to promotional material for cultural tours in the NT. The website also promised visitors that they could expect to "[l]ive for a day our *happy, creative & spiritual* way of *traditional* Aboriginal life" and undergo "[a] *life changing journey of discovery* in the Blue Mountains … Inspiring *authentic ancient* cultural; education & connection with adventure" (emphasis added).[4]

Visitors were thus being offered the capacity to Indigenise themselves through temporary access to an Aboriginal way of life as something they could themselves "live", seeming to suggest that visitors could experience what it means to be an Aboriginal person. A discourse of authenticity was a strident feature of this narrative, framed in terms of the time-depth of Aboriginal culture, its spirituality and the power of tradition. There was also a tendency to romanticise a traditional Aboriginal way of life in the Blue Mountains as "*happy, creative & spiritual*". Vicariously participating in this mythologised way of life, it was suggested, would lead to transformational personal change and discovery. The promotional material for this tour thus appeared to engage and perhaps reinforce the type of stereotypes about Indigeneity that Bunten (2018) and Carr et al. (2016) caution against in the Aboriginal cultural tourism context, to facilitate the appropriation of an Aboriginal identity by visitors.

The guide and business owner of Blue Mountains Walkabout explained in a later interview that he offered the tour to,

> … try and teach …[visitors] how culture used to be in this area before colonisation and also teach them the common Lore of the culture which was spread across Australia … And the other aspect is that I don't want

> it to be a history lesson. I want it to be relevant to modern day applications. I do teach it ... in ways they can practice it ... when they get home, actually incorporate ancestral ways of life in the modern world to get all the health and healing advantages that the ancestors had. So that to me is the most important part of the day.

Although he gave a brief account of the post-contact history of the Blue Mountains at the beginning of the walk, including the deaths of many Aboriginal people of this region from introduced diseases like smallpox, this, he explained, was "for necessity only", so that visitors "can understand what's happened to the local people, where are they all and why are they lighter skinned, and why they don't look like a typical Aboriginal, like a pure blood (sic) would look like, you know". The guide himself was light-skinned and blue-eyed, so in offering this context he may also have been trying to pre-empt questions about his own Aboriginality. The guide's comments indicate he was aware of the hegemonic settler-colonial discourse that an "authentic" Aboriginal person should look a certain way and that his guests might have certain expectations about him on this basis. However, he also capitalised on this discourse by promising guests the "discovery of [an] authentic ancient cultur[e]", in his web-based promotional material. I argue that such discourses oppress but also enmesh, so that it can be difficult for Aboriginal peoples working in contexts such as the tourism industry to disentangle themselves from these.

The guide later explained that he wanted to keep any content about the post-contact experiences of Aboriginal people to a minimum because,

> I wanna keep a sort of real happy vibe for the day. Because that's how the ancestors mostly used to live, in a state of contentment and joy and happiness of their day and that's what we're trying to replicate on the Walkabout, that experience, you know, for me to sort of get bogged down in what's happened post-colonial and what's happening with Aboriginal culture nowadays, it dilutes and potentially contradicts, you know, the whole experience.

The guide's assertions about the emotional state of the "ancestors" were for him deeply held beliefs, and related to his ideas about the health benefits of the pre-invasion lifeways of Aboriginal peoples. However, the conviction that prior to colonisation Aboriginal people lived in an unmitigated state of happiness and wellbeing engages the romantic "noble savage" discourse that has been critiqued by Indigenous scholars as a problematic stereotype that is applied to Indigenous peoples (Dodson 2003; Langton 1996). Discourses that seek to romanticise traditional Aboriginal lifeways are thus not confined to visitors but may also be employed by Aboriginal peoples in the cultural tourism context, and as Bunten (2018) has argued, this practice can serve to perpetuate these stereotypes.

Field journal recollections, Blue Mountains Walkabout tours, September, 2019

On the day of the tour our guide met us close to Faulconbridge Railway station in the lower-Mountains and took us to a small park, where the unsignposted walk commenced. We were a miscellaneous group of seven, mainly comprising international visitors from the United States. After a short introduction to the area and its post-contact history, we began walking through beautiful country, varying from dry Eucalypt forest with some impressively large Angophoras (Sydney Red Gums) higher in the walk, to some remnant cool climate rainforest along a stream in the valley. As a long-term Blue Mountains local I was amazed by the size of some of the trees, particularly the Angophoras, which our guide estimated could have been up to 250 years old. He encouraged us to touch the trees as we passed, engendering a form of sensory knowing (Pink 2015). One of the group described the Angophoras as "lumpy and cool" to the touch in a later interview, demonstrating how embodied affective experiences become inscribed within memory (Beck and Sommerville 2002; Wright et al. 2012).

Along the way our guide knowledgeably informed us of the names of the plants we passed and their uses by the Aboriginal peoples of this region. I learnt that a plant that grows naturally in my own garden, Lomandra, that I had considered a weedlike nuisance because of its spiky flower head, was used by local Aboriginal women for making bread. Since then, I've looked at this plant in a new way, and observed that the seed head is full in spring, which would have been when it was harvested. Lomandra continues to prompt a complex affective response for me, partly white shame and guilt, as I witness the seed going to waste, but also a feeling that I've established a closer connection to Country, coming to understand the plant in a different way, accepting it and looking after it by removing surrounding weeds, rather than objecting to its presence.

One of the highlights of the day was being shown a collection of rock engravings of wallabies on a large rock shelf with some water flow over parts of it, not far into the walk (Figure 4.4). These impressive engravings are located close to the urban interface and a major highway. Their only protection from the vandalism that other examples of Aboriginal material culture have encountered in this area is that their location is unsignposted, and presumably not widely known. Coming face-to-face with these again reminded me that I live on stolen, unceded Aboriginal Country, an uncomfortable but necessary truth. The sense of an Aboriginal presence that the engravings evoked made the physical absence of the people who made these markings more striking, calling to mind Gelder and Jacob's (1998) notion of the post-colonial uncanny.

We spent some time near the engravings, our guide explaining the meaning of some of the markings, and that engravings like this were only made by an initiated Aboriginal man of high degree. Visitors were entranced by

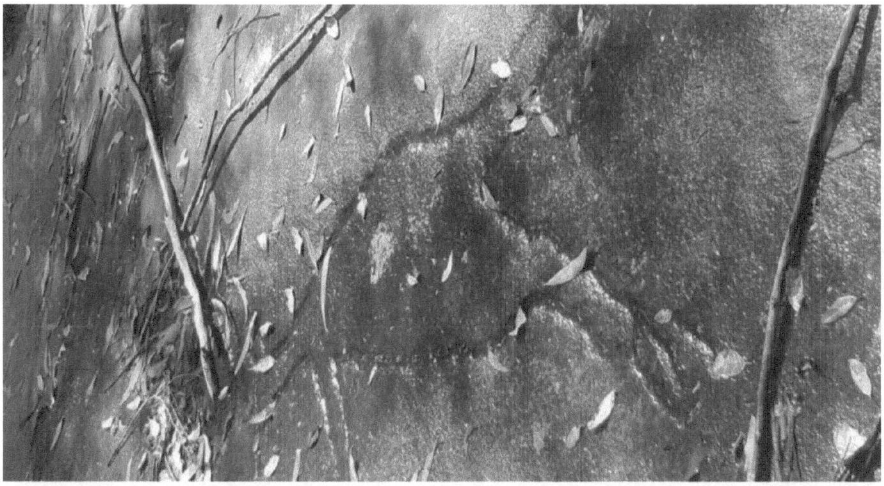

Figure 4.4 Wallaby rock engravings near Faulconbridge, Blue Mountains, NSW. Photograph by the author.

an Eastern Water Dragon sunning itself on a nearby rock shelf, reluctant to move when we came by, knowing it belonged and we were just passing *through (Figure 4.5). The high clear call of Eastern Whipbirds echoed around us, the sound vibrating through our bodies, joyful and uplifting, the non-human ever present.*

We paused for some time a little way from this place and our guide used the engravings to explain some aspects of Aboriginal spiritual beliefs about creation figures such as the Rainbow Serpent and Earth Mother and Sky Father (Baiamee), prompting visitor questions and comparisons with the beliefs of First Nations peoples of the United States. Although Indigenous cultures share similarities, it is their distinctiveness that is obscured by this type of dialogue, reflecting a settler-colonial tendency to collapse Indigenous peoples into a single category, a type of discourse our guide did nothing to correct.

The next stop was a sandstone rock overhang (Figure 4.6), where we had the opportunity to grind ochre, and if we wanted, to paint it on ourselves. Although some ground ochre no-one took up the option of wearing it. Being invited to grind and wear ochre was one way in which this cultural tour provided visitors with the opportunity to Indigenise themselves. "Signing onto Country" by making ochre handprints and stencils on trees at the Yellomundee Culture Camp, discussed below, was a similar practice. While relevant scholarship (d'Hauteserre 2015; Saul and Waterton 2018; Smith et al. 2012; Tucker and Shelton 2018) suggests that some visitors are seeking tactile, embodied tourist encounters, activities that involve Indigenisation of non-Aboriginal people cannot escape entanglement in a settler-colonial identity politics. This is because, as Veracini (2015, 61) explains, "in settler colonial contexts indigenisation (sic) was and is about the replacement of one

Figure 4.5 Eastern Water Dragon, near Faulconbridge, Blue Mountains, NSW. Photograph by the author.

socio-political collective with another". Settler Indigenisation through Indigenous cultural tourism is thus distinctly different in its meaning and effect to Indigenous peoples themselves reclaiming and remaking their own cultural practices (Clifford 2013), as this often functions to challenge the erasures of colonisation, rather than perpetuate them.

After having lunch under the rock ledge (Figure 4.6), we set off again down to the bottom of the valley, into a rainforest with a stream flowing through it, finishing at a waterfall, surrounded by a rounded rock amphitheatre that our guide referred as "the Earth Mother's womb". The final leg was to climb

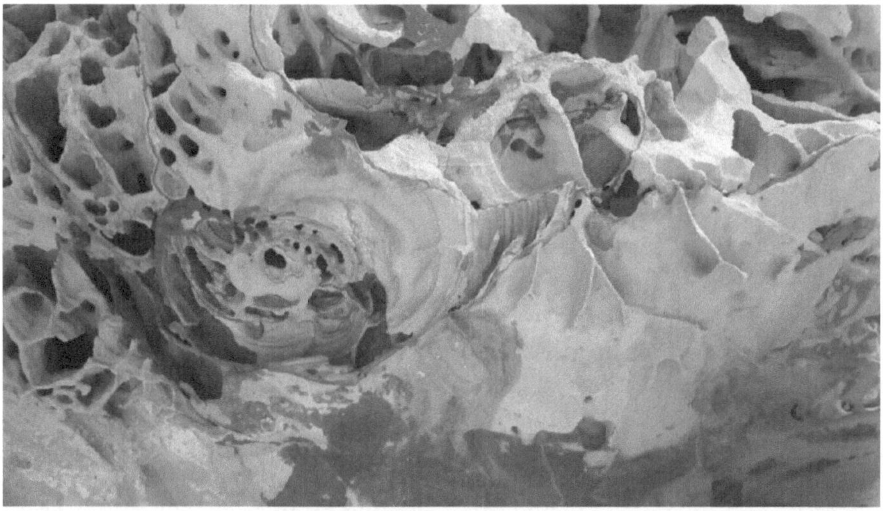

Figure 4.6 Sandstone rock formations, near Faulconbridge, Blue Mountains, NSW. Photograph by the author.

back up a steep hill to arrive at our starting point. I found this steep hill at the end of the journey exhausting, as did another woman, who was a little slower than me. Our guide clambered ahead like a mountain goat, later commenting on the difficulty of the hill, but simply observing "the Mother put it there".

The walk was beautiful, the engravings were impressive, and our guide seemed knowledgeable about the plants and animals of the area and the traditional spiritual beliefs of the Aboriginal peoples of NSW. However, he combined this with health advice to visitors about diet (what to eat), when to eat (adults not until noon), when to sleep (shortly after sundown), what to wear (least clothes as possible) which he called the Lore of the Dreaming. Rather than the guided tour format allowing our guide to tailor his offerings to suit his audience, one of the benefits that Hughes and Morrison-Saunders (2005) associate with this format, this guide appeared to have a pre-packaged message that he wanted to convey, and did so in a way that was quite didactic. My reading was that what was offered by Walkabout Tours had strong elements of a New Age spirituality, a perception reinforced by a later conversation with a Dharug Traditional Custodian about the tour.

In a similar way to the tour offered by Maraku Arts, the Blue Mountains Walkabout tour seemed to reify rather than challenge the existing preconceptions of visitors about Indigenous peoples (Bunten 2018; Carr et al. 2016), which do not reflect their complex, contemporary realities. However, the inclusion of elements of New Age philosophy in the tour did appear to reflect the guide's beliefs and identity and was a form of reinterpretation or remaking of Aboriginal lifeways that he thought would also have utility for his audience.

Yellomundee Culture Camp, Hawkesbury area, 2019

Yellomundee is a Regional Park on the western bank of the Nepean River within the Hawkesbury Local Government Area, on the eastern escarpment of the Blue Mountains. The Park is managed by the NSW NPWS and lies within the traditional territory of the Boorooberongal clan of the Dharug people. Aboriginal peoples assist the NSW NPWS in caring for Country at Yellomundee by weeding invasive species in the surrounding landscape and re-introducing traditional burning practices through a local Land Care Group (Darug Ngurra et al. 2019).

I first went to Yellomundee in August 2019 to consult with Uncle Lex Dadd,[5] a Dharug man from Western Sydney, about my desire to conduct interviews with visitors at Red Hands Cave in the Blue Mountains National Park, and also to interview him about the interpretive signage at Red Hands Cave. After finding out more about my research, Uncle Lex told me about the Culture Camps he runs at Yellomundee with academics from Macquarie University and suggested I include an upcoming camp as part of my research. This interested me as the Culture Camp was a prolonged immersive experience camping on Country and I thought visitor responses could be usefully compared to some of the shorter experiences discussed above.

Uncle Lex summed up the purpose of the camps as follows:

> We started up camps, they're called Yanama Budyari Gumada, which in my language means to 'Walk with Good Spirit', with patience, humility and respect, to create a safe place for people to come and visit and connect to Country. We crush ochre up and, just like you see with the hand stencils at Red Hands Cave, we get people to sign into Country, but we only do it on trees ... When they blow their handprint on that tree using the ochre in their mouth, that's their commitment to walk with good spirit, not only for Country, but for each other ... And we get in and have great cultural experiences. I'm a wood carver, we have a dance circle down here, and we have corroborees ... And we've run 22 successful camps here.

Another Uncle who helped with the camp I participated in said that his aim in offering the camps was "to give ... insight into what Aboriginal culture was and is now, and understanding of what the connection to Country is, and to share cultural information that people like Lex and I are able to share". He emphasised that what he wanted participants to take away with them was,

> ...[a] true appreciation of the significance of this Country, especially the connecting to Country. The smoking ceremony teaches people what that's about, and teaches people that we're surrounded by our spiritual ancestors, and going through the smoke is a sign of respect to them.

The smoking ceremony was thus one way in which the Culture Camp allowed participants to connect with an Aboriginal ontology of Country that included still present spiritual ancestors. Through the smoking ceremony, the use of ochre and the approach to Country modelled by the Uncles who led the camp, Aboriginal onto-epistemologies that emphasise relationality with the more-than-human and the agentic nature of Country were conveyed to participants, as they reflected in later interviews (see Chapter 7) (Bawaka Country et al. 2013; Palomino-Schalscha 2018). However, when asked what messages he wanted to convey to participants, the response of the Uncle quoted above was more overtly political:

> That our culture is strong and continuing and has always been a continuum, apart from all of the unfair and unjust legislation and what the government has done. They tried to destroy us, basically, and like any nation that conquers a race of people, it's always divide and conquer, and they did that by trying to weaken our core ceremonies, responsibilities and relationships with Country and each other ...

For this Uncle, the continuing practice of cultural ceremonies and responsibilities was thus one way to assert the strength of Aboriginal peoples and cultures to resist settler-colonialism, a discourse of rights also echoed by other Indigenous peoples in a cultural tourism context (Palomino-Schalscha 2018). The fact that cultural ceremonies now practiced in south-eastern Australia may have been revived, may not be the same as past practices and may represent a self-defined form of survivance does not undermine their validity.

Field journal recollections, September 2019

The first thing I noticed when I walked down to the camp set up by the river was how full of life it was, with many small birds darting and singing and the distinctive sound of Eastern Whipbirds calling, despite how overrun the landscape was with weeds. Many of these were huge vines that had climbed up and weighed down the Casuarina. Despite the weeds, the river itself was beautiful at this location, away from the town of Penrith that is my usual vista, crossing the bridge over the M4 motorway on my way to Sydney (Figure 4.7). I was surprised to encounter peace and beauty so close to the urban interface with Western Sydney. Like some of the camp participants I later interviewed, I too had dismissed this landscape as spoilt and not worth my time, without really bothering to engage with it.

Everything at the Culture Camp was set up when I arrived. The camp was laid out with chairs around a central fire that burned constantly and a kitchen tent where food was prepared, with other smaller sleeping tents located around the periphery (Figure 4.8). The surrounds of the camp fire became a lived-in space, as people claimed chairs by hanging their jackets over them.

Figure 4.7 View of the Nepean River from Yellomundee Culture Camp, NSW. Photograph by the author.

The camp formally commenced with a smoking ceremony, when Uncle Lex burned gum leaves and invited us to walk through the smoke as a sign of respect to the spirits of Aboriginal ancestors, still present at this place. After the ceremony, the perennial smell of the camp fire smoke, which imbued our clothes and hair, served to remind us that the ancestors are always present.

Figure 4.8 Chairs around the fire, Yellomundee Culture Camp, NSW. Photograph by the author.

Later on the first day we all shared in grinding white ochre in a wooden bowl, as we talked around the fire. I felt a little separate as I was not part of the student group, although everyone was friendly enough. A group went off with an Uncle to collect firewood, a regular activity to keep the fire going.

I saw a large white bird, a pelican or crane on the river. A cow was lowing in the distance. In the early afternoon, an eagle swooped down to the river with talons extended, parting the water, but caught nothing. I re-arranged a dentist appointment to be present for the "signing into Country" (see Uncle Lex's explanation of this activity above). Things unfolded slowly there, but I hoped this happened before I had to leave.

Before the "signing into Country", I weeded with the other participants, chatting and bonding as we stretched to grab and remove the vines in the Casuarina, carefully bagging them before they could drop their seeds. This weeding, mainly done standing, although an embodied practice engaging the whole body, also allowed time for conversation. I talked about the struggle with weeds at my own Blue Mountains property, which backs onto bushland, trying to prevent a variety of them from entering the bush. Just like the reclamation work at Yellomundee, this can feel like a never-ending task, that must be constantly repeated.

I also shared my sorrow over a neighbour's recent removal of a large Angophora because it apparently had a disease. We talked about whether the disease meant the tree really had to be removed, or could it perhaps have been left to die in its own time? My concern for Country where I live seemed connected to the weeding I was doing at Yellomundee, beginning to create a joined-up sense of Country in my mind and a series of affective interconnections.

The weeding had to be done carefully otherwise we risked dropping the small seed gourds of the balloon vines, which were prolific spreaders. The gourds themselves were delicately beautiful and each contained one round black seed – we tried to think of alternative uses – jewellery perhaps?

One girl was swimming in the river, exclaiming "Oh, Oh" as she entered the water. She was okay, just cold. Back around the fire, one of the group was talking about becoming a vegetarian because she wanted to care about the earth. The group were chatting, bonding. We were ready to "sign into Country".

In preparation for my own attempt, I watched two young women as they made hand stencils on trees (Figure 4.9). Taking a mouthful of water mixed with ochre, blowing and making raspberry noises led to laughter – but they made identifiable hand stencils. I discovered that I was not as talented, spitting not blowing: my handprint was a spit blob. I saw Uncle laughing discretely at me behind his hand. The young women kindly told me it takes a while to get it right.

Day two, one of the Uncles took us up to the dance circle, showed us the movements for emu, echidna, kangaroo and goanna. The wind came up in the tops of the tall gums as we danced, barefoot. The Uncle led us gently and slowly through the day's activities, seeming to know the right time for

Figure 4.9 Tree hand stencil, Yellomundee Culture Camp, NSW. Photograph by the author.

everything. I was conscious that I had to leave soon because of my dentist appointment: a somewhat harsh intrusion of normality. This meant that I couldn't really relax and just be in the moment and experience the type of "co-present immersion" Waterton (2015, 58) discusses in relation to another form of affective methodology, the "walk-along" interview. Awareness of my role as a participant observer didn't help. The time pressure, and my decision not to camp but visit, limited the extent that I could fully engage with the life of the camp. Perhaps it was the wrong decision. The following day was a day of cold and rain. I was at home, worrying about how the group was faring and called but couldn't get through.

The next day the river was up, the sun was trying to come out, the frogs were singing in the swamp. The pelican was on his stick in the river: I realised that it was his place. The students slowly came back to camp: most took a break and went home the previous day as the camp was awash due to the heavy rain. The young men helped Uncle Lex make spinifex glue – he used this to put together a stone axe. Much of the day was taken up by students individually interviewing Uncle Lex for their student projects. I realised I could have arrived much later. There was lots of swimming in the river that I thought was too dirty to swim in.

On the last day, Uncle Lex gave a "culture talk". We gathered together under the tarp around the smoking fire, eyes watering. Uncle Lex showed men's and women's tools, such as different types of boomerangs, digging sticks, coolamon, clapping sticks and talked about their various, often multiple uses. He pulled out a mime puppet he called his grandfather and told a story about a flying fox. The moral was that everyone's included. He joked with one of the students in a teasing but kind manner. We laughed, but not in a mocking way. There was an ethic of caring, not just for Country (Uncle Lex would say as Country), but for the people who were for that moment part of Country that was very strong in this small community, led and modelled by the Uncles. I made time to swim in the river. I was a little scared because it was running strongly after the rain and a current swept down one side. Immersion! It was cold but wonderful.

Driving home on the last day, I felt a sense of cultural dislocation. Five minutes after leaving the camp I was in the Blue Mountains town of Winmalee. It was about 4 pm, so the traffic was busy. I felt that I'd left a small enclave that operated by different rules, an Aboriginal ethic of caring and community, and come back to the "real" world. It was jarring and peculiar, as I left one way of being to return to another. But when I arrived home I had a strong insight that I live on Country. I knew this before, but I saw it afresh, or felt it more deeply. There were some of the same birds at home as at Yellomundee, some of the same trees. The light was similar, although heavy rain fell at home today and it was cooler. Differences, similarities and connections traced across the landscape linking people and places, both to each other and Country, not simplistically, or without grief, loss and pain, but there was also joy, humour and affinity, and the sense of growing into something, of co-becoming, together.

Discussion

Leaving the world of the river behind with some reluctance, but still aware of Country as I notice the small birds darting in the garden outside my study window, I seek to make sense of the Yellomundee Culture Camp and of the other cultural tour experiences described in this chapter, through the process of writing about them. Although the literature suggests that one of the advantages of the guided tour over static forms of interpretation such as interpretative signage is the dynamic two-way interaction of the guide with the visitor (Wearing and Neil 1999), which allows messages to be shaped to the needs of visitors (Hughes and Morrison-Saunders 2005), it was only the Yellomundee Culture Camp that fully demonstrated this type of interaction, as there was an openness and flexibility to the format that facilitated it. A true interchange between guide and visitor is thus perhaps not something automatically generated by the guided format but one that must be enabled by the guide, by allowing space and time for these interactions to occur. Guided tours in which the guide was motivated to offer the tour because they wanted to communicate a specific message, and did so in a didactic way, such as the Blue Mountains Walkabout Tour, appeared to allow less opportunity to productively engage the visitors. The Maraku Arts format in which the interpreter acted as an intermediary between the guide and the visitor, also limited the interaction between the two. All of the guided experiences, did, however, allow two-way interaction through the opportunity visitors had to question the guides, and many took advantage of this. Visitors were thus to some extent able to explore specific issues that interested them by asking questions of the guide, an opportunity that was not present when engaging with static, passive forms of interpretation such as the on-site signage discussed in Chapter 3.

Moving from the guided tour format to consider the interpretative content presented in the tours, a commonality shared by all of the Traditional Custodian led cultural experiences described in this chapter was the emphasis placed on the continuity of Aboriginal cultures in place. This was achieved by presenting Aboriginal cultures as arising in "deep time" yet with aspects that continue into the present. In one case, revived cultural practices such as smoking ceremony in south-eastern Australia, were identified as a way for Aboriginal people to resist the impact of colonisation, and thus as a form of survivance. However, only one cultural experience, the Patji tour, chose to fully integrate a discussion of post-contact Aboriginal history into the tour content, and thus presented political and social events which continue to impact the contemporary lives of Aboriginal peoples as suitable subject matter for a cultural tour. All the tours also presented Aboriginal peoples as having an important relationship with Country, demonstrated through connections with protected areas rather than urban landscapes (although Yellomundee is on the urban fringe). A further commonality was a shared tendency to present Aboriginal cultures as monolithic and as closely related to tangible objects that were part of pre-colonial material culture (men's and women's tools, for example), rather than arising in the contemporary social worlds of Aboriginal peoples in all of their complex manifestations.

However, if culture is understood in the terms explained by Macdonald (2001) in Chapter 1, it is about dynamic, changing contemporary practices, and as a consequence, a focus on tradition can only tell part of the story of culture. It will inevitably fail to illuminate current social aspects of culture that arise in part from the impacts of settler-colonialism upon Aboriginal peoples, including "the nation's laws, practices and ideology in relation to Aboriginality" (Cowlishaw 2009, 181). As Gillian Cowlishaw argues, Aboriginal identity is "linked to embeddedness in an Aboriginal social world" (Cowlishaw 2009, 188). I argue that cultural experiences such as that offered by Maraku Arts and the Blue Mountains Walkabout Tour gave limited insight into the current social worlds of Aboriginal peoples. The absence of any contemporary social or historical content in the Maraku Arts Cultural Tour, or indeed any indication of the way that cultural practices have changed over time, gave an impression of Aṉangu culture as static and timeless, serving to perpetuate an authenticity discourse about Aboriginality.

Although I argue that the cultural tours I participated in had a tendency to present Aboriginal cultures in a simplistic way, rather than as entangled in complex social worlds, this was more the case for some tours than others. As noted above, Blue Mountains Walkabout Tours tended to portray a simplified, romanticised view of Aboriginal cultures that represented Aboriginal peoples as part of nature. This is problematic because equating Indigenous peoples with nature has historically been a rationale for Eurocentric philosophical, social and political systems to deny Indigenous peoples their full humanity as part of a racialised colonial discourse (Byrne et al. 2013; Lee 2016; Wearing and Huyskens 2001). Although Aboriginal peoples have a relationship with Country that they are born into (Moreton-Robinson 2022), recognising this is not the same as reducing Aboriginal peoples to a part of the "natural" world and thereby reinforcing a nature/culture dualism.

The association of Aboriginal peoples with nature, sustainability and communality apparent at the Yellomundee Culture Camp, acted as a further powerful discourse of authenticity which could result in Aboriginal peoples who do not conform to these expectations being framed within a deficit discourse of deculturation (Paradies 2006). Although the relationship with Country remains important to many Aboriginal peoples, including those in urban areas, (Grieves 2008), presenting this as a totalising identity mythologises Aboriginality in ways that do not necessarily accord with the diverse lived experience of all Aboriginal peoples today (Paradies 2006). The forced removal of many Aboriginal people from their Country, particularly in south-eastern Australia, which bore the early brunt of colonisation, means that many Aboriginal peoples no longer live a traditional lifestyle on Country, or continue to reside on their traditional lands. However, this does not invalidate their Aboriginality.

While I do not wish to criticise movements of cultural revival such as the Yellomundee Culture Camp, the belief that traditional Aboriginal culture was more communally orientated, less regimented and more environmentally sustainable than Western culture only makes sense from the vantage point of the present. In other words, the Aboriginal past was very much reinterpreted, of necessity, through the lens of the present day. In so doing, a place is also found for settler-colonial Australians such as myself to fulfil their desire for belonging by suggesting that because we live on Country we can all care *as* part of Country, and "that humans and Country exist together in a state of co-becoming" (Bawaka Country et al. 2013, 195). The adherence to a philosophy that we are all part of Country, as espoused in the concept of caring *as* Country by Bawaka Country et al. (2013), subsequently becomes a rationale for the invention of rituals such as "signing into Country", that mirror former Aboriginal ways of interacting with Country, such as the creation of hand prints and stencils at Red Hands Cave.

This was enacted at Yellomundee Culture Camp as a way of affirming the bond that settlers can also experience with Country, based on the philosophy that we are also a part of Country. However, a number of authors argue that the seeking out of Aboriginal cultures, spiritual practices and the claiming of Aboriginal material heritage as the heritage of all Australians can be a type of cultural appropriation designed to legitimate settler claims to belonging (Byrne 1996; Lattas 1992; Mackey 1998; Muir 2014; Waldron and Newton 2012). Patrick Wolfe (1999), for example, contends that a logic of elimination governs the way that Australia as a settler-colonial nation responds to Aboriginal peoples, whether by massacre or assimilation, leaving their cultures free to be fully appropriated in their absence. However, the perspectives of two Indigenous scholars working in the field of critical Indigenous studies, Aileen Moreton-Robinson (2022) and Margaret Mutu (2022), are distinctly different, in that they contend that Indigenous cultures are not available to other peoples to claim, regardless of the presence or absence of Indigenous peoples. Moreton-Robinson explains that the Aboriginal ontology of Country is encompassed by Aboriginal sovereignty, which is incommensurable because it is "inherent" and "inextricably connected to being in and of" Indigenous lands (Moreton-Robinson 2022, 259). Similarly, Mutu bluntly states that Māori sovereignty is "inalienable so it cannot be ceded or given away to others" (Mutu 2022, 269).

The extent to which non-Aboriginal people can share in the ontology of Country is thus necessarily limited by the fact that such people are not "in and of" Aboriginal lands in the same way as Aboriginal and Torres Strait Islander peoples. It is also important to query the more-than-human concept of co-becoming and to point out that our different socio-political positionings and personal and collective histories mean that Aboriginal and non-Aboriginal people do not co-become in the same ways. In a similar vein, Sarah Radcliffe (2017, 225) critiques the way that more-than-human geographies seek out "Indigenous peoples to illustrate non-western MTH [more-than-human] engagements, enlivening geographical imaginations even as they downplay … colonial-modern and settler geographies".

A comparable invitation to share culture and enact forms of visitor Indigenisation was not one that typified the Patji Tour on the outskirts of Uluṟu-Kata Tjuṯa National Park. The Patji cultural tour shared aspects of culture in accordance with Tjukurpa, to teach visitors the norms and rules that they were expected to comply with while on Aṉangu land, including the limits on knowledge and behaviour. Visitors were thus very much seen as outsiders who may not understand how to behave in the "proper way" on this Country if this was not explained to them, separated from Aṉangu as the custodians of this land and holders of both knowledge and responsibility. While visitors were encompassed by the laws that apply to Country when they were on Country, they were not considered part of Country in the same way as Aṉangu. While Aṉangu want visitors to understand their relationship with Country, its stories and take the place away "in their hearts" (Breeden 1994, 190), cultural boundaries are maintained and not all aspects of culture are shared.

Part of the important information that our Aṉangu host on the Patji tour wanted to share with visitors also concerned his family Tjukurpa, and this narrative covered a range of confronting issues including state sponsored frontier violence, the Stolen Generations, Aboriginal dispossession and the land rights movement. It thus dealt with the type of settler-colonial impacts that Cowlishaw argues have influenced the contemporary identities of Aboriginal peoples. The inclusion of this material indicates that these are subjects that the Uluṟu family, a Traditional Custodian family with responsibilities towards Uluṟu, considered important for visitors to know and understand as part of their encounter with Uluṟu-Kata Tjuṯa National Park and surrounding Country. Given the highly moral nature of Tjukurpa and the fact that visitors are encompassed by it, acquiring this type of knowledge could be considered a moral duty for a visitor. The Uluṟu family, facilitated by the small group tour provider, were thus seeking to construct a visitor experience that focused not just on traditional knowledge and beliefs, but demanded a level of political and moral engagement from the visitor by articulating the unpalatable truths of settler-colonial history. This agenda is similar to that of Bawaka Cultural Enterprises in North East Arnhem Land, where visitors might be told about the impacts of the Northern Territory Intervention on the community, and asked why they did nothing to stop it (Lloyd et al. 2015).

However, an eagerness to broach such topics was not apparent in my consultations with staff employed by authorities responsible for the management of natural areas. For example, when asked about whether the inclusion of interpretation on the Stolen Generations would be an appropriate subject for park-based interpretation within Uluṟu-Kata Tjuṯa National Park, a Parks Australia manager responded,

> that is more of a social thing. And yes, that was an impact [on Aṉangu], but if you're going to do that there are so many other terrible things

> that happened, you'll have to do all of them. So that's just one and the (NT) Intervention being another. So, I'm not keen for that.

He went on to explain,

> I think it can be mentioned, but I don't think it needs to be dwelt upon ... this isn't the place for a discourse about some of the social issues and injustices of the past, I think they can be mentioned, they shouldn't be hidden. But I certainly don't believe they need ... to be dwelt on in a significant way.

The comment that a national park such as Uluṟu-Kata Tjuṯa is not "the place for a discourse about some of the social issues and injustices of the past" seems based on the assumption that injustice towards Aṉangu is confined to the past, and therefore lacking in contemporary relevance. The discourse that I argue is favoured over "social issues" in protected area interpretation, is that of Western science, particularly that of environmental conservation and natural resource management (Howitt et al. 2013; Lee 2016), and what I have termed an authenticity discourse about Aboriginal peoples. Despite Uluṟu-Kata Tjuṯa National Park's joint management arrangement, it appears that a dissident Aṉangu voice speaking about frontier violence, nuclear missile tests, the Stolen Generations and the land rights movement could only be heard in an Aṉangu led cultural tour, which ironically also took place outside of national park boundaries. The linkages that exist between post-contact history, Aṉangu and the natural environment of the park, although stark, have not been able to broach the nature/culture divide that characterises park-based interpretive discourses.

In Traditional Custodian led cultural tours, particularly those offered by Aboriginal businesses, there are no bureaucratic constraints on content and for some tour providers there is a willingness to introduce topics that could be challenging for a settler audience. Yet this greater freedom does not necessarily mean that in all cases an alternate postcolonial history or contemporary understanding of culture is promoted, due in part to a desire to meet perceived visitor expectations of a pleasant experience and to fulfil their desire for an encounter with "authentic" Aboriginal culture (Bunten 2018). However, my research suggests that some visitors were not averse to having key aspects of post-contact history addressed in cultural tours and other forms of interpretation in protected areas, in fact, when asked, several indicated that they would welcome it. As Vera, a retired domestic Australian visitor who participated in the Maraku Arts Cultural Tour, explained,

> I think we have to [address these issues]. I think we have to finally turn around and face it ... Finally. In the early days "We didn't invade!", but yes we did. It offended us because we didn't want to hear it but now that we hear it and understand it, yes we did.

Carrie, a young British woman living and working in Australia responded similarly,

> Oh yeah, I'd love to know more about that ... Yeah, I think you know, people need to be told, if people don't know the truth then how's a wound going to heal? ... And once people know the truth, they can have more of an understanding.

Conclusion

I thus conclude that Indigenous cultural tourism such as the Patji tour has a role to play in Makarrata or truth-speaking in the postcolonial Australian context. Not only this, by turning the tourist encounter to their own ends, and offering "hot" interpretative content that could result in negative emotions for visitors, Patji decolonises the visitor encounters with Aboriginal cultural tourism. This tour also foregrounds a discourse of rights seen in other Indigenous tourism contexts that has been found to enhance the status of Indigenous peoples (Palomino-Schalscha 2018; Theodossopoulos 2018). On the other hand, experiences such as Blue Mountains Walkabout Tours reinscribed what I argue is a damaging authenticity discourse about Aboriginal peoples and cultures, by romanticising traditional Aboriginal lifeways. The theme of the authenticity discourse is one I take up in the next chapter, where I introduce narratives arising from visitor interviews that bring into sharp focus the problematic and ongoing nature of this discourse as it is applied to Aboriginal peoples and cultures in protected areas.

Notes

1 The Uluṟu monolith is referred to colloquially, mainly by domestic Australian visitors, as "the rock".
2 A number of major and minor nuclear tests were conducted at Maralinga and other Australian sites from 1956 to 1963 by the British Government, with the agreement of the Australian Government, https://www.nma.gov.au/defining-moments/resources/maralinga, accessed 26/09/2022.
3 Intrepid travel website, https://www.intrepidtravel.com/au/australia/uluru-adventure-127138, accessed 20/01/2022.
4 Blue Mountains walkabout website, https://www.bluemountainswalkabout.com/frames.htm, accessed 20/01/2022.
5 "An 'Elder' is an identified and respected member of an Aboriginal and Torres Strait Islander community. Elders generally hold key community knowledge and are expected to provide advice and support to community members. Age alone does not necessarily distinguish an Elder. Many Aboriginal and Torres Strait Islander peoples acknowledge Elders as 'Aunty' or 'Uncle' as a sign of respect, even if they are not related" (Queensland University of Technology 2021, 3). Lex Dadd was introduced to participants in the Yellomundee Culture Camp as "Uncle", and so was referred to as such by participants, including myself.

5 Authenticities, deficits, "Aboriginalisms"

Visitor constructions of Aboriginality in protected areas

Introduction

This chapter explores the motivations, expectations and experiences of visitors when they travel to places of recognised cultural significance to Aboriginal peoples within protected areas, and how this relates not only to the stereotypical beliefs about Aboriginality that some visitors bring with them but also to their own positionalities, predominantly as white Australians. Although I am also concerned with the ways in which the attitudes and perceptions of visitors changed as a result of their experiences, these changes are the focus of Chapter 7. The current chapter concentrates instead on the pervasiveness of certain settler-colonial discourses about Aboriginality, primarily within Australia, and how these also appear in a tourism context. The arguments that I will make are thus in line with those of Erickson (2018, 43), who contends that "in touristic encounters with race … the differences encountered (and deemed an ethically valuable feature of tourism) are structured by a discourse of whiteness even before the tour begins".

Arising directly from the accounts of visitors themselves, I identify the concept of authenticity as significant in this regard. I use this term to refer to the settler-colonial belief that there are real or genuine Aboriginal people, and another group of Aboriginal people who cannot be regarded as such because they no longer live a traditional lifestyle on Country. The characteristics of this supposed "authentic" Aboriginality include time-depth, environmental sustainability and living in harmony with Country in parts of Australia classified as remote rather than urban. Not only was this authenticity discourse revealed in the protected area interpretation already discussed but in visitor responses to the places they visited and the Aboriginal peoples that they engaged with.

Underlying some visitor responses was also a white paternalism that prompted certain categorisations of Aboriginal peoples, based on the assumption that their status as white Australians authorised them to do so. In this sense, as Erickson (2018, 44) argues, whiteness is more than a subject position or classificatory discourse, but "an organizing discourse for the way that we see race itself … [and] is privileged as the pure original position of

DOI: 10.4324/9781032654973-5

the human subject". I counterpose a racialised concept of authenticity, which I link with an ongoing politics of surveillance and control of Aboriginal peoples founded in white privilege, with a type of experiential authenticity that visitors often seek through tourism, associated with affective, embodied touristic encounters in place. A third form of authenticity is also present in visitor discourse: one that seeks an Aboriginal cultural tourism experience that is considered reliable as it is authored or authorised by Aboriginal peoples themselves. In discussing these varying types of authenticity, I draw on literature from the fields of critical Indigenous studies, decolonial and settler-colonial studies, tourism studies and contemporary anthropology, in which these understandings of authenticity both converge and diverge.

Referring back to this scholarship, I argue that the belief that certain groups of Aboriginal peoples are more "authentic" than others arises from the association of geographic remoteness from urban centres and closeness to an ideal of "traditional" Aboriginality with the original and genuine. This resembles a form of "object" authenticity in its belief that a primordial Aboriginal authenticity is "accessible in its true form" (Belhassen and Caton 2006, 855). Discourses of authenticity as applied to Indigenous peoples have been substantially critiqued by a number of Indigenous and non-Indigenous scholars working in critical Indigenous studies and related fields (Dodson 2003; Fredericks 2013; Gorringe et al. 2011; Grieves 2008; Lambert-Pennington 2012; Maddison 2013; Moreton-Robinson 2016; Porter 2018). As these authors note, beliefs about timeless, romanticised "native others" are accompanied by a necessary parallel discourse that associates urbanised Aboriginal people with deficit or loss, an equally pervasive and deleterious way in which settler-colonial subjects categorise Aboriginal and Torres Strait Islander peoples.

However, it is not only a deficit of culture that some Indigenous Australians are assigned: epithets of disadvantage, lack and abject victimhood are part of the typical discourses used to describe Aboriginal peoples in both institutional and popular discourses. Because of its purchase "across different sites of representation, policy and expression", Cressida Fforde et al. (2013, 162) identify deficit discourse about Aboriginal peoples, after Foucault, as exhibiting all the characteristics of a discursive formation. My research adds weight to this argument. The Australian Government policy of Closing the Gap is one example of an institutional approach that frames Aboriginal and Torres Strait Islander peoples in this way, due to its representation of all aspects of Aboriginal identity, from health status to educational attainment, in terms of deficits (Fforde et al. 2013; Fredericks 2013; Gorringe et al. 2011).

While I am not arguing for a direct, deterministic relationship between what visitors say and the institutionalised language of deficits apparent in government policy approaches, media discourses, educational settings and elsewhere, obvious corollaries exist. It is through a deficit discourse that Aboriginal peoples continue to be recreated as the "other" of the white, Western subject, a form of "Aboriginalism" (Hodge and Mishra 1990, 27) that resembles the "Orientalism" applied to peoples of the fictive ensemble known as "the East" by nineteenth-century colonialism (Said 1978). Although contemporary Aboriginal and Torres Strait Islander peoples such as trawlwulwuy woman Emma Lee reject the status of "other" that they are ascribed, it is nevertheless important to understand the ways that settler-colonial discursive structures perpetuate this othering (see Chapter 2).

In categorising Aboriginal peoples via discourses of authenticity and deficit, some visitors also assumed that they had the capacity and the right to define what it means to be Aboriginal, and on this basis, whiteness operates as a relational category and the norm against which all peoples are judged. Whiteness is thus "about the discursive practices that because of colonialism and neocolonialism, privilege and sustain the global dominance of white imperial subjects and Eurocentric worldviews" (Shome 1999, 108–109). In making this argument, I do not mean to imply that all visitors are white, nor that all white visitors thought in this way, but that the perspective described was significant among the visitors I encountered.

Within this context, notions of authenticity operate as an "oppressive" ideology of the settler-colonial state, in which the racialised concept of "indigenous (sic) purity" is deployed when "varying degrees of biological and cultural authenticity [are required] before [settler-colonial governments will grant] political recognition and economic support to indigenous (sic) people" (Sissons 2005, 39). Indigenous peoples are thus continually required to "authenticate their Indigeneity for access to welfare, citizenship, or Native title" (Moreton-Robinson 2016, 115). Far from innocuous, the racialised ideas about Aboriginal cultures and cultural identity articulated by some visitors are implicated in historic and ongoing efforts to define and control Aboriginal peoples and for this reason, ideas of authenticity must be understood as historically, socially, politically and culturally situated (Silverman 2015).

However, as a number of authors in the field of heritage, tourism studies and anthropology argue, authenticity is a relative concept and can manifest in several different ways, even in the same touristic encounter (de Bernardi 2019; Martin 2010; Mkono 2013; Silverman 2015). While I identify the search for the primordial Indigenous "other", linked with the belief that there is such a thing as "object" authenticity, as a quest for authenticity that is "oppressive" (Sissons 2005, 39), other ways in which the concept of authenticity manifests are more benign, or less easy to categorise. For example, and as will be discussed, some visitors articulated a desire for Indigenous cultural tourism to be delivered by Indigenous rather than non-Indigenous people, and that a Traditional Custodian standpoint be clearly represented in interpretive signage dealing with their cultural heritage. For other visitors, clear information indicating that audio-tours on mobile phone applications had been developed and authorised by Traditional Custodians was necessary in order for them to be considered reliable.

Far from being oppressive, such perspectives reflect an implicit acceptance of Indigenous sovereignty and the right of Indigenous peoples to be the authorities on their own identities and to control how they are represented (Dodson 2003). Other forms of authenticity, as they appear in the tourism literature and are reflected in visitors' experiences, cannot be simply categorised as one thing or the other. Most notably, visitors' desire for what has been labelled "existential" authenticity (Wang 1999), whereby they engage in processes of being and learning through direct embodied engagement with place and its human and non-human others (Knudsen and Waade 2010; Rickly-Boyd 2013), may serve to challenge rather than reinforce oppressive versions of authenticity, depending on the nature of this engagement.

In the remainder of this chapter, I explore the various manifestations of authenticity outlined above, as they appear in visitor discourses, in relation to the literature in this field. The chapter is structured thematically rather than by research locale, although, in the case of the first sub-heading, the themes identified arose specifically in visitor discourses associated with Uluṟu-Kata Tjuṯa National Park. The section that follows discusses the way that visitors to the various research locales romanticise Aboriginality and how this functions as a form of "oppressive authenticity" (Sissons 2005, 39). The final section draws again on visitor accounts to argue for the rehabilitation of the idea of authenticity as it applies to the direct, embodied experience of place and Aboriginal cultures some visitors are seeking.

On "fixing the Aboriginal problem": Deficit discourses and paternalism in Uluṟu-Kata Tjuṯa National Park

Underlying the discourses of authenticity and inauthenticity introduced above is a form of white paternalism that assumes it is the prerogative of settler-colonials to judge, assess and classify Aboriginal peoples from a presumed position of superiority, and, where judged necessary, intervene "to fix" perceived problems. This is part of a process of racial stereotyping described by Stuart Hall (1997, 258), whereby the more powerful social group "divides the normal and acceptable from the abnormal and the unacceptable [and] then *excludes* or *expels* everything which does not fit, which is different" (emphasis in original). Fundamental to this process of expulsion, is the tendency of white colonisers to subject Aboriginal and Torres Strait Islander peoples, as Yawuru scholar Michael Dodson (2003, 32) observes, to a "continual flow of commentary and classification". Wakaya scholar Yin Paradies (2006, 355) explains this classifying tendency as a "need to surveil and control the socialization, mobility and biological reproduction of those with some descent from pre-colonial peoples of Australia". Classificatory discourses using notions of blood quantum as the basis for institutionalised forms of control of Aboriginal peoples at the state and commonwealth level such as policies of protection and assimilation were, in fact, key ways that settler-colonialism was established and maintained in the Australian context (Carlson 2016).

Discourses about Aboriginality arising in visitor interviews varied considerably based on whether interviews were conducted in Uluṟu-Kata Tjuṯa

National Park or in protected areas in NSW within driving distance of Sydney. In the mid to lower Blue Mountains on Sydney's urban fringe, Aboriginal cultures tended to be treated as relic and Aboriginal peoples as absent. However, in Uluṟu-Kata Tjuṯa National Park, the knowledge that A<u>n</u>angu continue to live in the park and maintain their languages and aspects of their traditional cultural practices, while lauded by some visitors as more "authentic" than their urban counterparts, also prompted a range of judgements and concerns.

For example, I encountered Angus, originally a Scottish national but at the time of the interview a long-term Australian resident of Victoria, at the former climb site in May 2019 after he had climbed Uluṟu. Angus was dismissive that A<u>n</u>angu have any *a priori* claim to Uluṟu that gave them a greater right to it than any other group, musing,

> Oh, I don't know if anybody can *own* anything, if you know what I mean. And even if they call themselves the Traditional Owners, I think either there was nobody or someone here before them. And I don't think that gives you a right to ownership.

Later in the interview, he reflected on the "hand-back" video playing at the Uluṟu Cultural Centre, noting that "they [A<u>n</u>angu] were happy to get it [the national park] handed back, but I don't know, necessarily, if they've done anything with that, since the handback". Angus not only denied the significant relationship that A<u>n</u>angu have with Uluṟu as Traditional Custodians but he also immediately assumed a position of white superiority, taking it upon himself to judge whether A<u>n</u>angu have made good use of the park "handback", simply on the basis that he is white. In this way, "whiteness is constituted in forms of epistemic privilege" (Moreton-Robinson 2003, 66) as the power to know and judge the adequacy of Aboriginal peoples, regardless of their own sovereign understandings of self that tebrakunna country and Lee (2022, 149) refer to as a "self-located place of belief and knowing".

Angus was unconsciously adopting the perspective that whiteness is normative (Erickson 2018), and that the behaviour of non-white "others" should be judged based on this norm, which is treated as natural and not requiring explanation or justification (Hall 1997; Moreton-Robinson 2003; Shome 1999). However, whiteness is value laden, not value neutral, incorporating a number of unquestioned assumptions including that land is a resource to be utilised, to do things "with" or to. This is a Eurocentric way of understanding land, quite removed from Aboriginal ontologies of Country, that empower Aboriginal peoples to care for Country rather than act upon it (Howitt et al. 2013). Not only did Angus assume that a white ontology was the norm and that it should operate within Uluṟu-Kata Tjuṯa National Park but his discourse also infantilised A<u>n</u>angu, harking back to the colonial belief in the child-like native satirised by Said (1978), as one who could neither govern themselves nor make good use of what the beneficent white state bestowed upon them. Angus' response suggests that visitors' engagement with places such as Uluṟu-Kata Tjuṯa National Park is mediated by their own assumptions and beliefs about Aboriginal Australians, based on racialised discourses in wide circulation in the Australian social, political and cultural context (Dunn et al. 2004; Lentin 2020; Moran 2009).

Other visitors, such as Carlos, originally from Chile but also a long-term Australian resident, demonstrated a variation on these views, suggesting that migrants to Australia from culturally diverse backgrounds can also adopt postcolonising "white" norms. Interviewed on the Liru walk close to Uluṟu, Carlos expressed concern that A<u>n</u>angu were "isolated from everyone else", and that the rest of Australia was "moving ahead", but A<u>n</u>angu were not. The assumption that Aboriginal communities who are far removed from major urban centres are remote and isolated is based on adopting the perspective of the city dweller as the norm, when Aboriginal communities in these locations are living on their traditional lands, in relationship with Country, as they have done for generations. Furthermore, the belief that the rest of Australia is moving ahead but A<u>n</u>angu are static assumes that because A<u>n</u>angu culture has ties with tradition, it is not also dynamic and adaptive.

As Duelke argues, the present may be "built on the persistence of tradition ... while at the same time, serving to accommodate within itself the inevitability of both change and dynamism..." (Duelke et al. 2005, 201). However, this nuanced understanding of culture is not a feature of "colonial/modernity" (Quijano 2007, 176), a term used by decolonial scholar Quijano to convey that modernity operates as a form of colonialism, enforcing conformance to its values and modes of being among societies judged to be pre-modern. A<u>n</u>angu are thus obliged to abandon their traditional beliefs and lifeways, perhaps even leave their Country so that they are no longer "isolated", as they proceed along the linear road of progress to modern futures (Tuck et al. 2013).

Without actually having met any A<u>n</u>angu or after only limited interactions with A<u>n</u>angu working in the tourism industry, a number of other visitors also assumed that A<u>n</u>angu were "backward", or had problems that they, the non-Aboriginal visitor, were somehow responsible for remedying. While this concern was genuine, these questions might not have arisen had they visited a predominantly white "outback" community. For example, Bill, an older domestic Australian participant in the ranger guided walk at the base of Uluṟu, discussed his desire to improve the situation of Aboriginal peoples who have been negatively impacted by white colonisation. He explained,

> ... you know, it's good, the way the Aboriginal (sic) has, you know, been respected more [recently] because since the white man got here, he was just downtrodden, downtrodden. And, you know, basically we invaded their land and it was totally wrong .... We can't sort of change that, but we can at least try to make amends to make a better life for them, we've got to ask them, what do they want? We can't just say, well, here's a house or here's this, because unless you're actually taught what it's used for, as the ranger just said, you know, they'll rip the doors off, take the windows out.[1]

Bill summed up the invasion of Australia by Europeans in moral absolutes, as "totally wrong", and although he stopped short of expressing feelings of guilt or shame, he articulated a desire to "make amends" and in this way acknowledged a form of collective settler-colonial culpability. These views identify him as an Australian who had a partial foundation for establishing more progressive relations with Aboriginal and Torres Strait Islander peoples. However, he simultaneously constructed Aboriginal peoples as needing the help of white people "to make a better life for them". In so doing, Bill did not seem to realise that there is an irresolvable contradiction in the perpetrators of this "wrong" unilaterally appointing themselves the ones to remedy it. Echoing the guide on the ranger guided walk, he also repeated the familiar white trope that Aboriginal peoples do not look after what they are "given" by a beneficent welfare state, because they allegedly damage government provided housing by "rip[ping] the doors off".

The phrase "ripping the doors off" seemed to operate as an emotionally charged collective symbol for settler visitors, evoking fears of social chaos and Aboriginal intractability, as part of a deficit discourse about Aboriginal peoples (Fforde et al. 2013). This phrase also evokes the supposed "otherness" of Aboriginal peoples based on their difference from a Eurocentric norm (Hall 1997; Hodge and Mishra 1990; Said 1978). However, anthropologist Gillian Cowlishaw (1999; 266, cited in Byrne 2003b, 176) explains that removing doors from houses is not a marker of social chaos but could facilitate a different form of sociality to that based on a white model. Altering housing in this way could aid community interaction by enabling Aboriginal peoples to "watch ... people as they come and go ... be available to kinspeople, and communicat[e] directly with finger talk". While removing doors from houses may serve to facilitate Aboriginal sociality, interpreting these actions simply as senseless property damage allows Aboriginal peoples to be constructed as a problem, but one that can now be dismissed as too hard for governments and concerned individuals to solve. In this way, settlers can be absolved from any further responsibility to "make amends".

Ray and Naomi, two domestic Australian visitors from coastal NSW who travelled to Uluṟu with their young children, also positioned Aboriginal peoples as a problem to be "fixed" by Australian Governments and one for which they somehow felt responsible. Like Bill, they demonstrated an awareness of the negative impact of white invasion on Aboriginal peoples, but tended to characterise this as producing a range of deficits and dysfunctions, as evident in the interview excerpt below:

Ray: I think the general perception that white people have of the Aboriginals (sic) isn't good, they look at all the problems that they have, and the social issues which are totally being caused by Europeans coming here and they kind of just judge them and base everything around that. And I think it's really quite derogatory to them as a culture.

Naomi: But it's also very hard to fix, isn't it? The government is trying to fix things, and money doesn't fix things, and housing doesn't fix things. It's a very hard issue.

Ray: Like the guy [ranger] was saying today, they build them houses and stuff like that. It's not what they want. I don't know what they want, but I've heard many times that they build them houses, they go in and they pull them apart and they use furniture for fires ... I'm a plumber by trade and I've worked in a few of the Housing Commissions around [coastal NSW town] ... that have ... Aboriginal people living there and the living conditions. Like it's a real shock. ... a lot of poverty, a lot of lack of education. There's social issues, drug and alcohol issues. But yeah, and it's, I'm the first person to put my hand up and say I'm extremely ignorant. I don't know what they want, and I don't know how to fix it.

Ray and Naomi echoed common popular and governmental discourses about Aboriginal peoples as socio-economically disadvantaged and experiencing a range of social problems such as drug and alcohol misuse, which they traced back to the impact of white colonisation. They thus reproduce the type of classificatory commentary that Dodson (2003) argues at the beginning of this section is applied to Aboriginal peoples. Engaging the colonial fiction of the "degenerate native" (Langton 1996, 11), Ray assumes in the above excerpt that alcohol abuse is a problem for Aboriginal peoples. This assumption is based on a generalised stereotype that operates as not only individual but shared [false] cultural knowledge, creating expectations and norms about how Aboriginal peoples will behave (Balvin and Kashima 2012). Such views were most starkly apparent in the discourse of Pete and Penny, also domestic Australian visitors; however, unlike Ray and Naomi, they attributed this supposed deficit to the failings of Aboriginal peoples themselves.

Pete, whom I interviewed immediately after he had climbed Uluṟu, explained his lack of interest in the Aboriginal heritage of the park by asserting "We have 'em at home and they're the worst of the worst". When asked to explain what he meant by this statement, Pete launched into a racist diatribe, referring to Aboriginal peoples as a monolithic "they" and claiming, "They do not look for any employment, they don't, they just don't. It's hard to describe, when they don't respect 'emselves how are we going to respect them too?". Pete did not distinguish between Aboriginal peoples from his home state of Queensland and those living in the NT, except to claim, "They're far drunker here to be blunt and honest, they are far drunker here".

However, it soon became apparent that Pete had had no contact with A̱nangu on which to base this judgement, and his claims of drunkenness related to Aboriginal peoples he had encountered in other towns in the NT. In fact, Mutitjulu is a dry community and alcohol can only be purchased by visitors and people who have a residence permit because they live and work in the park. The only people who can be observed drinking alcohol in the park or Yulara are visitors, regardless of the inappropriateness of bringing alcohol into a dry community and drinking it around Aboriginal peoples who are denied the same privilege.

While Pete's views about Aboriginal peoples were based on his own deep racial prejudice, in attributing deficits to Aboriginal peoples that needed to be "fixed", Ray and Naomi were engaging in a similar discourse, but from a different perspective. As canvassed in the introduction, various commentators have labelled a deficit discourse "a narrative of deficiency" (Fforde et al. 2013, 162), which defines Aboriginal identity in terms of lack. As Fogarty et al. (2018, vi) explain, the term deficit discourse "particularly denotes discourse that narrowly situates responsibility for problems with the affected individuals or communities, overlooking the larger socio-economic structures in which they are embedded". I argue that one characteristic of this discursive formation, as evinced in commentary by Bill, Ray and Naomi, is the positioning of Aboriginal peoples as a problem for settler-colonials "to fix", at the same time that this problem is presented as one that is unamenable to solutions. Not only this, there is also a tendency to assume that Aboriginal peoples are not themselves future-orientated, and that they will remain fixed in this problematic state until white people intervene to rescue them (Bandyopadhyay and Patil, 2017).

While Bill, Ray and Naomi, unlike Pete, did not hold such "problems" to be the responsibility of Aboriginal peoples, but linked them with the impact of white invasion, they nevertheless subscribed to a belief in Aboriginal disadvantage as a totalising identity, leaving no room for notions of survival and resilience. One signal of this was their tendency to refer to Aboriginal peoples as "the Aboriginal" or "the Aboriginals", evoking an archetypal notion of Aboriginal peoples as one distinct, monolithic group, rather than a diverse range of peoples. Ray and Naomi also expressed a level of anxiety around the belief that they, as relatively privileged white Australians, were somehow responsible for fixing what they perceived as the intractable problems of Aboriginal peoples, Ray concluding in evident frustration "I don't know what they want, and I don't know how to fix it". These anxieties represented a common affective response for visitors (Grimwood et al. 2019; Slater, 2019), and suggest that Uluṟu-Kata Tjuṯa National Park is a "moral terrain … inscribed onto bodies through affect" (Waitt et al., 2007, 249) (see Chapter 6).

Such beliefs were not in keeping with the way that senior A<u>n</u>angu such as Nellie Patterson describe their culture and wish to promote it to visitors, Nellie affirming,

> We want to hold our culture strong, so we can teach and learn from generation to generation. When visitors talk to A<u>n</u>angu, 'Oh, I say, A<u>n</u>angu really have strong culture and there's big Tjukurpa here'. And for men and women there's lots of sacred and powerful things. We are teaching young people lore and culture and learning to the ones who come up after us.

Nellie does not present Anangu as "other", but emphasises their self-defined ways of being and knowing (tebrakunna country and Lee 2022). In my interview with her, she also chose to focus on the positive work Anangu are doing to care for Country and to keep lore and culture alive and although she made it clear that Anangu aim to work in partnership with Parks Australia, they are clearly not awaiting white rescue.

> **"We'll always be climbing!", May 2019, UKTNP**
>
> *It was Sunday and my day off, so I decided to do the difficult Valley of the Winds walk at Kata Tjuta, with my partner who was visiting from the Blue Mountains.*
>
> *There was a short steep rock climb with the unfortunate name "Little Uluru" that formed part of the walk. In fact, you had to climb over "Little Uluru" to continue.*
>
> *A woman ahead of me gesticulated widely, throwing out her arms while standing on the top of "Little Uluru", as she shouted "We'll always be climbing!".*
>
> *I considered letting her know I didn't share her views, but couldn't catch her up.*

Falling from grace: Noble savages, "degenerate native[s]" and "anomalous" others

The deficit discourse examined in the previous section was an important underlying narrative that came into play when some visitors attempted to articulate their response to Uluru-Kata Tjuta National Park, Anangu and Aboriginal peoples more generally. However, a prominent rationale for the decision to make the long and expensive journey to the park in the first place was a desire to experience not only what was perceived as an "authentic" Aboriginality but also what some visitors described as the "real" Australia, a type of "object" authenticity quest that is very much socially constructed (Belhassen and Caton 2006; Wang 1999). This quest for the "real" Australia is at least in part a fiction promoted by the tourism industry to create a range of iconic destinations that will appeal to domestic and international audiences, one that hinges around the romanticisation of white Australia as a pioneer nation (Smith 2012) and of places such as Uluru-Kata Tjuta National Park as the spiritual heart of Australia (Pomering and White 2011). This imaginary of the spiritual fixes upon "romantic" geographic sites that are seen as infused with authentic Aboriginality and which are associated with a deep past (Byrne 1996), places that non-Indigenous people now have access to through heritage and tourism.

A number of scholars link the search for an authentic Aboriginality to a romanticisation of pre-colonial Aboriginal peoples as akin to Rousseau's "noble savages" (Dodson 2003; Hinkson and Fullenwieder 2019; Langton 1996). Travel to remote regions as a quest for this version of Aboriginality is derided by Dodson (2003, 33) as a search for "the noble, well-built native: heroic, bearded, loin-clothed, one foot up, vigilant, with boomerang at the ready". As Marcia Langton (1996, 11) reveals, this "caricature" is inevitably paired with that of its opposite, the "degenerate native", with both terms functioning as disparate but mutually dependent foils within white constructions of Aboriginal identity. However, Dodson (2003, 33) also identifies a third, more recently emerging, genre of romanticised Aboriginality as "ochred, spiritual, and playing the didgeridoo behind the heroic travels of a black Landcruiser". In this final category, the heroic noble savage is replaced by the settler-colonial in their "heroic" travels across inland Australia, with Aboriginal peoples again romanticised as spiritual, but with their identities reduced to a musical accompaniment to white adventures in "becoming Aboriginal" (Galliford 2012). This demonstrates how, as anthropologist Jeffrey Sissons (2005, 9) has argued, "[n]arratives of settler nationhood routinely employ indigenous (sic) imagery to create romanticized pasts that ... [have] been transcended or succeeded" by settler subjects.

The decolonial historian and theorist Patrick Wolfe (1999, 173) argues that the romantic noble savage genre of the authentic Aboriginal "is important to national ideology because the parallel coexistence that it depicts is consistent with the legitimising illusion that Australia was not founded on homicide and theft". At the same time, "the vicious genre furnished a justification for these foundations, which were ideologically insulated within the liminal space of the frontier". According to Wolfe (1999, 163), such fictions of the good "authentic" and bad decultured Aboriginal person both serve the objectives of the settler-colonial state. He coined the term "repressive authenticity" to describe an "official way of talking about ... Indigenous people" (Wolfe 1999, 179) in response "to the threat posed by a multiplex, heterogenous and, above all, *historical* set of Aboriginalities that refuse to be contained within the ideal polarity that the logic of elimination requires" (Wolfe 1999, 190, emphasis in original).

Sissons (2005, 39) is similarly critical of what he labels "oppressive authenticity", by which he means settler-colonial definitions of Indigenous peoples in terms of ideas of purity and impurity, based in part on the underlying eugenics of blood quantum, which function "primarily as a mechanism of exclusion; those who cannot be placed securely within one of two categories – 'native' or 'settler' – become people out of place". While Sissons does not identify the operation of oppressive authenticity as part of a "logic of elimination" by a hegemonic state actor, or emphasise the historically specific nature of Aboriginality, the critique of authenticity by both scholars is congruent. I use visitor accounts to argue that repressive/oppressive authenticity does not manifest only in official discourses but also in the ways that tourists understand and articulate their own experiences of engaging with Aboriginal peoples and places. Several Indigenous people I interviewed who worked in protected areas and the cultural tourism industry also identified being on the receiving end of white discourses of oppressive authenticity in their day-to-day working lives as an ongoing burden. For example, a light-skinned Indigenous Ranger who worked at Yellomundee Regional Park explained that people are "disappointed when I turn up to meetings as the Aboriginal representative" because he did not conform with their expectations of how an Aboriginal person should appear.

Desire for an authentic Aboriginality is also part of a romanticisation of Aboriginal peoples and places as spiritual destinations as part of a settler desire to claim this perceived spirituality for themselves. This romanticisation is apparent in tourist advertising of locations like Uluṟu-Kata Tjuṯa National Park, which invites visitors to "explore the great Australian desert and feel the overwhelming presence of the majestic rock. Discover Australia's *spiritual heart*" (emphasis added).[2] Although most visitors I spoke to did not believe their travel decisions were directly influenced by such advertising, some nonetheless framed their reasons for travel in similar terms. However, visitors also had unique and highly personal reasons for making their journey, including celebrations of important life events, highlighting their often multiple rationales for travel. This was the case for Naomi and Ray, introduced above, who travelled to Uluṟu-Kata Tjuṯa National Park with their young family in May 2019.

Framing her journey as a form of existential self-actualisation through engagement with an iconic landscape (Wang 1999), Naomi explained,

> I've wanted to come here for a long time to have a deeper sense of understanding of the country that I was born in. I've always had this sense of Uluṟu's being like the heart of our country, and it was always a real spiritual journey to come here.

Later in the interview Naomi further elaborated that one of her goals in visiting the park was also to "bring my children here, to educate them about a world bigger than their lives that they live in the moment and to expose them to a *different* culture and a *different* reality of how people live" (emphasis added). The emphasis on Indigenous alterity in this statement by Naomi is striking, evoking Erickson's (2018, 44) understanding of whiteness as "an organizing discourse for the way that we see race itself", where whiteness is "privileged as the pure original position of the human subject", and all who are not white are therefore labelled "different". But Naomi and Ray were disappointed and conflicted about their lack of contact with Aṉangu during their time in the park, Ray noting that they had not encountered any Aboriginal staff within the Uluṟu Cultural Centre. This lack of contact led them to question whether Aṉangu want to engage with visitors. In this context, Naomi mentioned her primary school-aged children "asking … the whole time we've been here 'We want to meet the Aboriginal people! Why can't we talk to them?'"

Naomi, Ray and their children thus came to the park wishing to engage with a "different culture" characterised by what they perceived as an "authentic" Aboriginality they assumed was absent from the coastal region of NSW where they lived, although Ray's earlier comments reveal that he had encountered Aboriginal peoples through his job as a plumber. While their children's request "to meet the Aboriginal people" could be dismissed as an objectifying desire for contact with the exotic "other", it was to some degree aligned with the desire of Anangu to share their culture with children that came through quite strongly in an interview with two senior male custodians. On this basis, the desire of Naomi, Ray and their children to meet and interact with Anangu was not out of keeping with the sort of engagement with visitors that some Anangu were also open to, however, mechanisms did not exist to facilitate this.

Similarly, Frank, interviewed on the Lungkata walk at the base of Uluru, was clear that his desire to travel to the park "was something that was always on our bucket list for a long, long time". He explained that his reason for travelling at this time was to celebrate his 47th wedding anniversary. Nevertheless, the language Frank used to describe his desire to visit Uluru echoed some of the discourse of the above advertisement, referring to it as a place that is,

> ... not just iconic but a place that had a sense of *spiritual connection* to our land and, and Uluru being one of the least colonised places, given that, you know, regular contact between the Anangu people and the white settlers didn't really begin until the 30s and 40s, you know, in any big way, and it seemed to be a place that was ... absolutely Australian in its feel and we wanted to be here. (Emphasis added)

For Frank, the appeal of Uluru-Kata Tjuta National Park related to his belief in its object authenticity as a spiritual place (Belhassen and Caton 2006), although he did not fully explain what he meant by this concept. He also considered the park to be quintessentially Australian, perhaps reflecting the success of tourism advertising campaigns in constructing Uluru as an iconic Australian destination for both domestic and international visitors (Pomering and White 2011). Frank justified his assessment of the park as a spiritual place by referring to it as a less colonised environment than other parts of Australia because white incursion occurred later there. In Frank's account, white colonisation was thus perceived to have a tainting influence on Country, as if it was in a state of original purity prior to this event and this essence had subsequently been lost. This assessment has parallels with the way that Andrew, a young university student who participated in the Yellomundee Culture camp, described the Yellomundee landscape as "disturbed", "degraded" and "not untouched by a long way" because it had been completely covered in invasive weeds before the Landcare group, led by Uncle Lex, began work to reclaim it.

But Andrew found that his attitude to this landscape changed during the Culture Camp, explaining,

> ... one thing that's sort of struck me is like valuing what you would call degraded landscapes, because previously, I was sort of really into, you know, going out to what you would call wilderness, I guess, even though it's not really a thing, but I guess, as most as untouched as possible ... And so that's meant for me sort of overlooking what you would call like disturbed or degraded landscapes and sort of writing them off as ... a bit too far gone, or like they've just been damaged too much. But yeah, coming here has, I think, shifted that for me, ...Yeah, that's sort of turned it around for me and gone well, you can't just give up on a place ... because we've damaged it.

While acknowledging that wilderness is a type of social construct with no basis in reality (Vidon et al. 2018), Andrew was nevertheless drawn to engage with places thus defined. However, he demonstrates a degree of reflexivity as a consequence of the Yellomundee Culture Camp by critically reflecting on how his desire for wilderness led him to dismiss what he terms "degraded landscapes" on the urban fringe, articulating a revised perspective that non-Indigenous people such as himself have a "responsibility to help that place", or care for Country. In saying this, Andrew is reflecting the perspective of the Traditional Custodians who led the Culture Camp that non-Indigenous Australians are also part of Country and can share in the responsibility of "caring as Country" (Darug Ngurra et al. 2019, 279). Through his experience at Yellomundee, Andrew thus underwent a shift in perspective that allowed him to experience a landscape on the urban fringe in different terms, reframing it as Country rather than as simply a place that has been "messed up" or "damaged".

As with landscapes, Aboriginal peoples could also be perceived by visitors as "fall[ing] from grace" (Dodson 2003, 33) after losing their Edenic innocence following white invasion. Jason, a retired domestic visitor first encountered while participating in a Maraku Arts cultural tour in Uluṟu-Kata Tjuṯa National Park, agreed to participate in a follow-up interview three months after our initial meeting to discuss his most important memories and any changes he had experienced following this event. He explained that the significance of the experience for him related to his:

> ... exposure to the *real* Aborigine[sic] [as] ... a mind opening experience rather than the urban ones. I lived in Glebe in my early years, I then lived in Chippendale, close to Redfern. I knew so many Aborigines (sic), kids and adults and, and to be honest, it was a matter of them looking for their next drink ... the *real*, *uncorrupted* Aborigine (sic) is a totally different creature to the urban one. They've been poisoned by white civilization: the alcohol, the drugs, petrol. (Emphasis added)

Jason counterposed what he viewed as "the real Aborigine" (sic), who was an Aboriginal person living remotely like those he encountered during the cultural tour he participated in, with Aboriginal peoples living in urban areas like inner Sydney, whom he regarded as "corrupted" by alcohol and other drugs. In Jason's case, his search for "the real Aborigine" (sic) through cultural tourism, appeared to arise from a desire for a "primordialist authenticity" (Theodossopoulos 2013b, 398–9) based on locating Aboriginal peoples within a "static conceptualization of the past". Although most Aboriginal peoples now live in regional towns and cities, in Jason's view, they are somehow less "real" than those living in remote parts of Australia. Evoking the way that Andrew formerly perceived "degraded" landscapes, Jason regarded urbanised Aboriginal peoples to have permanently lost some type of original essence they previously possessed, and like the colonised landscape, to be significantly damaged. But as Sissons (2005, 37) queries "[w]hy should first peoples (sic) be expected to have authentic identities while settlers and their descendants remain largely untroubled by their own ill-defined cultural characteristics?". Expectations that Aboriginal peoples should exhibit authentic identities are based on the same classifying logic of surveillance and control that underlies the deficit discourses discussed above and demonstrate the type of oppressive authenticity Sissons (2005) also critiques.

This contrast between the "authentic" Aboriginal person who lives close to nature and the perception that urbanised Aboriginal people are not only degraded but dangerous made a further appearance in my discussions with Vicki, another domestic visitor who participated in the ranger guided walk in Uluru-Kata Tjuta National Park. She reflected appreciatively on the willingness of Traditional Custodians in national parks throughout the NT "to share their knowledge and to want to share the place". However, this contrasted with Aboriginal peoples in,

> ... some of the other communities, possibly where they have *lost* a lot of their links to their land and the culture. It's a very different experience, you know, you go to some of the towns like Tennant Creek, which is quite scary. Compared to a place like this, you know, you sort of felt like 'I don't even want to walk down the street', sort of scary (emphasis added).

Vicki's standpoint as a white woman is reflected here in her perception of town-dwelling Aboriginal peoples as presenting a potential risk to her safety, a common stereotype applied to Aboriginal peoples, often men, in these settings. What Vicki is articulating here is the postcolonising binarism that distinguishes the culturally "authentic", "good" Aboriginal person living on Country from the decultured "bad" Aboriginal person living in towns or cities, who has, as the myth goes, been degraded by contact with white society and "lost" their culture. While the former prompted the emotion of gratitude, the latter produced the emotion of fear, perhaps of unpredictable violence, associated as places such as Tennant Creek are with myths of the frontier (Wolfe 1999). It is a manifestation of whiteness that Vicki felt empowered to comment on a place such as Tennant Creek, without understanding the history of virulent racism directed at Aboriginal people living in this town by white locals and a long history of neglect of the needs of remote Aboriginal communities by Australian governments (Altman 2018; Ashenden 2010).

An oppressive invocation of authenticity was also apparent in the romanticised assumptions that Eve and Katrine, two international visitors who participated in the Blue Mountains Walkabout Tour, had of Aboriginal culture, partly in response to the type of experience offered by the tour guide (see Chapter 4). Eve, a visitor from the United States, expressed her surprise that "you would find like the drawings on the rocks and things. I thought most of that was out in the outback. So, I had no idea it was as close to Sydney as we are". Eve was discussing engravings of wallabies on a sandstone rock that forms part of a watercourse we were shown on the Blue Mountains Walkabout Tour. She was not expecting to find such clear evidence of Aboriginal occupation close to Sydney because she assumed this could only be found in the "outback", a colloquial term used to describe Australia's remote desert regions. As discussed above, "authentic" Aboriginality equals remoteness, and the association of Aboriginality with an urbanised place not far from the invasion's epicentre was thus considered "anomalous" (Sissons 2005, 38; see also, Fredericks 2013; Porter 2010).

When Katrine, a French national, was asked to explain what stood out for her about the same experience she responded:

> I would say the mythology, the storytelling, with five phases of education or life ..., when I heard what he was saying, it was very similar to other native (sic) stories I've heard, you know, on other continents, very much in relation to nature and progression of life with nature.

For Katrine, the experience provided on the Blue Mountains Walkabout Tour reinforced her existing association of Indigenous peoples with nature, suggesting that they have an essence that remains premodern. This is not to say that a connection with Country, a concept that encompasses much more than the Eurocentric idea of nature, is not important to many contemporary Aboriginal peoples, including those based in urban areas (Grieves 2008). However, the identities of contemporary Aboriginal and Torres Strait Islander peoples are diverse and are to some extent socio-historically produced, including as a result of the postcolonising policies of Australian governments. For this reason, Paradies, argues for "free[ing] Indigeneity from the prison of romanticization" (Paradies 2006, 363) as opposed to an "essentialized Indigeneity [that] ... coalesces around specific fantasies of exclusivity, cultural alterity, marginality ... leav[ing] an increasing number of Indigenous people vulnerable to accusations of inauthenticity" (Paradies 2006, 355).

The question of waste, May 2019, UKTNP

Living in Rangerville while I do my fieldwork, I've never been so conscious of the waste that I, as a sole person, accumulate. There's no garbage service here and waste must be personally taken to a garbage dump at Rangerville or into town. As I don't have a 4WD, accessing the dump is difficult. So I must make surreptitious trips to dump my rubbish at Yulara.

There is no paper recycling, only bottles and plastics and cans. Most of my waste is food packaging and no choice about buying it.

Don't know what the garbage collection arrangement is in Mutitjulu, but there's lots of plastic waste lying around the roadways and wrecks of cars. This stands out in a national park. But I question my own expectations about pristine park landscapes, remembering that this place was an Aboriginal homeland before it was a national park.

A further manifestation of romanticism in visitors' discourses across a range of national park settings was in the description of "traditional" Aboriginal cultures as more sustainable than contemporary Western forms of social and economic organisation. For example, Eric, a young Irish national living and working in Australia interviewed after participating in a bus tour in Uluṟu-Kata Tjuṯa National Park, was powerfully impacted by the message that Australian Aboriginal peoples are:

> ... the oldest Indigenous civilisation and if that's the case they've obviously got something right, in terms of how you only take what you need ... and you can live sustainably off the land by being responsible ... I think that ultimately there's a lot that we [non-Aboriginal people] can take from their way of life. ... because ... you know I've been told that 98% of Australian rainforest has been destroyed since European settlement, and I think that just shows the impact we're having on our own planet.

Key features of a sustainability discourse as it appeared in the narratives of Eric and other visitors was that Indigenous peoples live in harmony with nature, utilising natural resources to meet their needs without destroying or overusing them. In contrast, post-industrialised societies of the Anthropocene were perceived as courting environmental disasters through deforestation and climate change. While attributing a positive value to traditional Indigenous lifeways, this discourse at the same time positioned Aboriginal peoples as closer to nature, running the risk that Indigenous peoples may be labelled "ecologically inauthentic" (Sissons 2005, 40) if they did not demonstrate these links in ways expected by settlers. Sissons (2005), for example, argues that in order to be considered "authentic" such links must be performed in overt and potentially stereotypical ways. It was, perhaps, a reflection of my own internalisation of such stereotypes that I was surprised to see widespread plastic waste and car wrecks on the outskirts of Mutitjulu during my 2019 stay in Rangerville, as discussed in the journal excerpt.

Chandler and Reid (2019, 14) critique what has been labelled as the "Ontological turn" (Todd 2016) in Western scholarship, which they sum up as the demand that "we 'become indigenous (sic)' in order to cope with and adapt to the new conditions of the Anthropocene". They critique contemporary philosophical approaches such as actor-network theory and more-than-humanisms for appropriating Indigenous ontologies without engaging with the work of Indigenous scholars or with Indigenous social and political realities, such as struggles over sovereignty. Not only is the concept of sustainability outside of Indigenous ontologies, which place more emphasis on relationality, utilising Indigenous lifeways as exemplars of what humans need to do to survive the current planetary ecological crisis aims to create viable settler futures on unceded Aboriginal land. While settlers may understandably wish to live in a less destructive relationship with their environments, this does not mean that Indigenous lifeways can be unreflectively deployed and (mis)represented as strategies for securing settler futures.

Non-hegemonic authenticities? Seeking a "living experience" and non "codified knowledge"

The deep entanglement of visitor encounters with Aboriginal peoples and places with oppressive authenticity might cause us to question, with Phoebe Everingham et al. (2021, 2), whether it is possible to do "tourism otherwise to colonial logics". There is, in fact, an emerging call to decolonise not only tourism scholarship but visitor engagement with Aboriginal peoples and the tourism industry, in recognition of the ongoing colonialism within tourism and academia (Chambers and Buzinde 2015; Grimwood et al. 2019; Stinson et al. 2021). However, I agree with anthropologist Keir Martin (2010, 339) that, "[a]ssertions of authenticity may be hegemonic from particular perspectives, but to presume that these perspectives exhaust the possibilities inherent in the concept is surely to deny the multiple variety of nuances of meaning dependent upon context inherent in any concept ...".

Consequently, I argue that the touristic desires and experiences in Uluru-Kata Tjuta National Park and the other protected areas discussed herein cannot be wholly encapsulated by the ideas of oppressive or repressive authenticity but also need to be understood in other ways, including as forms of embodied cognition (Knudsen and Waade 2010; Rickly-Boyd 2013). Accordingly, some visitors reported that interpretive signage at the base of the Uluru monolith created a strong feeling of connection with Anangu and their culture. Although such connections were necessarily superficial, in contrast with the relationships established on extended cultural tours such as the

Yellomundee cultural camp, it is a marker of the success of the Uluru-based signage that some visitors experienced a subjective feeling of connection to Anangu as a result of reading it. As reflected in the visitor narratives that follow, this is related to the fact that much of this text speaks about Anangu culture in the present tense, as a living continuing culture, addressing the visitor in both Pitjantjatjara and English.

Discussing the signage at the Mutitjulu waterhole, Tessa and Howard, two retired domestic visitors from Sydney, noted the use of the Anangu first-person. When asked about their response to that, Howard explained that for him,

> ... it emphasises the, what would you call it? Cultural capital, cultural centrality, ... ownership, knowledge, yeah, I mean it just emphasises that for me, that this ... is coming from a very deep personal space, it's not, this isn't *codified knowledge* that's being transmitted to us, this experience that's being told to us, [is a] *living* experience, which is, yeah, which is what we need. (Emphasis added)

Like a number of other visitors I spoke to, Howard was seeking interpretation that was personal, direct and arising from living Anangu culture, not abstract formal "codified knowledge". This suggests that some visitors do not want interpretation that is straightforwardly educational, as Staiff (2014) argues is the goal of most protected area interpretation, but interpretation that enables them to connect with culture and Country on an emotional level, by capturing "living experience". Howard's statements could be regarded as expressing a desire for authenticity in the tourism encounter with Indigenous culture, without at the same time assuming that he will discover Indigenous alterity (Paradies 2006). Similarly, Ben, a young man who travelled to Uluru-Kata Tjuta National Park to celebrate his 21st birthday, interviewed on the Lungkata walk, reported that the use of the Anangu first person in the signage he had read on this walk, "feels like it's a message directly from Country as opposed to national parks signage, which can be quite factual". He also reflected that in the absence of contact with Anangu during his time in the park, "it's been really good to gain that insight through anecdotal pieces".

Fran and Suzie, two young women who had grown up in the Northern Territory, one in the Mutitjulu community, also noticed and appreciated the Anangu first-person voice in the signage. Fran responded, "Yeah I like it, [use of Anangu first-person] I don't like to use the word *authentic*, but it feels like it's presented more *authentically*, versus from the perspective of just pure Parks signage" (emphasis added). While authenticity is a vexed concept when discussing Aboriginal peoples and cultures, Fran used the word to indicate that presenting interpretive signage in an Anangu "voice" felt closer to the spoken voice of Anangu than signage which simply reflected the discourse position of a bureaucratic agency such as Parks Australia. The reflective and critical way that Fran deploys the concept of authenticity, as June and Alex do below, is quite different from the way this concept appeared in the discourse of visitors such as Vicki, Jason and Ray because it is not associated with stereotypical preconceptions of an unchanging Aboriginality (Theodossopoulos 2013a, b).

This non-oppressive form of authenticity is well illustrated by Alex and June's discussion of what, for them, constituted an "authentic" experience of Indigenous cultural tourism, one that involved hearing a Traditional Custodian speak about Indigenous culture from an "insider" perspective. In both Fran's narrative above and June's below, the concept of authenticity is introduced by the interview participants, rather than the interviewer, reflecting the ongoing importance of this idea for Western visitors, particularly in the context of Indigenous cultural tourism (Mkono 2012, 2013). Alex and June, two self-styled "grey nomads" interviewed after they participated in the ranger guided walk, were not making judgements about Anangu in the dialogue below, but trying to describe the type of guided experience that, for them, was more direct and truthful, and therefore more engaging. They found being presented with information about locating bush foods by an Anangu guide to be a more meaningful experience than being told about Anangu lifeways by a non-Aboriginal guide during the ranger guided walk. Alex, comparing the ranger walk with the guided garden walk offered at a Yulara hotel, explained,

> His story [the Anangu guide's] was his story. He owns it. And when he was talking about the witchetty tree and the witchetty grubs and how if, if you see the signs on the ground, then you know that it's time, you can harvest them and they taste delicious, and 'Who's eaten them?' … And so it was a very personal engagement. Whereas this morning, [the ranger] was telling a story that he'd learnt.

June agreed, that when the Anangu guide

> … was talking about it, it's got that extra *authenticity*, because it's 'I'm talking about my stuff'. He owns it, it's his stuff. [The ranger guided walk guide] was great. He's very well informed. He's talking about their stuff …
>
> But [the Anangu guide's] talking the whole time about this. And, and he said, 'You know, … my little niece, she had a bucket of witchetty grubs, and I went over and she said to me, 'Don't you touch one'. And so he dramatises it … And that makes it just that much more *authentic* … And for me, that was, that was more fun. (Emphasis added)

For Alex and June, the guided garden walk was engaging because it was delivered by the person for whom it had the most relevance, as it was about their own culture, personal experiences and family. June and Alex's dialogue touches on the issue of what visitors consider an authentic experience, a subject of some contention within the tourism literature (Knudsen and Waade 2010; Martin 2010; Theodossopoulos 2013a, b). While some authors argue that because there is no agreed understanding of authenticity in tourism studies it should be abandoned (Steiner and Reisinger 2006), others contend that because the idea holds traction for visitors and the tourism industry, as evidenced in the narratives above, it is one that tourism scholars must continue to engage with (Belhassen and Caton 2006; Mkono 2012, 2013). In her discussion of Western visitors to Zimbabwe, Mkono (2013, 205), demonstrates that "authenticity, whatever it means to them, is very much part of the Western psyche in cultural tourism contexts". In fact, Belhassen and Caton (2006, 854) argue that "slippery concepts like authenticity" can be "highly functional" in the multidisciplinary social research discourse of tourism studies. Recent work by de Bernardi (2019), Deepak (2021) and Lovell and Hitchmough (2020) suggests that the concept of authenticity continues to hold traction for tourism scholars.

However, the relevance of the concept of authenticity to Indigenous cultural tourism, particularly in the Australian context, is an issue that has so far received insufficient critical scholarly attention, not only from the perspective of visitors but hosts, including Aboriginal cultural tour providers. For visitors, hosts and tour providers, this encompasses the issue of authentication; for example, the question of who has the right to speak about Indigenous cultures and under what circumstances, an issue raised by Finegan (2019) in the context of his discussion of protected area interpretation. In the context of Aboriginal cultural tourism, it is also important to consider the perspectives of international visitors, who may lack even a basic familiarity with Australian Aboriginal cultures. As the accounts of international visitors Helga and Laura illustrate, this lack of familiarity can result in questioning the authenticity of a tourism product that was taken as self-evident by domestic visitors, when it is not directly delivered by Traditional Custodians.

Laura from Brazil and Helga from Sweden had been in Australia for a short time when they took the Healing Waters Tour on the Jenolan Caves mobile phone app. Discussing their experience several days after this event, they expressed a strong preference for a tour provided in person by a Traditional Custodian over the Jenolan Caves' self-guided tour using the mobile phone app. Laura reflected that she would have paid more attention to tour content if it had been delivered by a guide, and that her preference would be for a Traditional Custodian guide because "… I would love to feel more connected. And … it's a person who, who is inside this culture, so this person would have more legitimacy to explain". While June and Alex preferred the insider perspective because it was more engaging, in Laura's view, a Traditional Custodian guide had more legitimacy to interpret the culture. Although the narrator of the Healing Waters tour was in fact a Traditional Custodian, Helga and Laura were unsure if this was the case as these details were not clearly explained within the app. Helga and Laura thus had questions about who designed and wrote the content. Helga explained that this led them to question, "could we trust this information or not? We had doubts the whole time". Not only does this response highlight the potential difficulty international visitors may encounter with self-guided tours on a mobile app, it also indicates that some visitors are seeking interpretation of Aboriginal cultures by people they believe can authoritatively and legitimately do so, because they belong to the culture they are interpreting (Salazar and Graburn 2014; Zhang et al. 2018).

Helga and Laura's desire for reliability of information and Alex and June's concept of authenticity in a guided tour are dissimilar to the perspectives of visitors who engaged with Aboriginal cultural tourism seeking a type of "primordialist authenticity" (Theodossopoulos 2013b, 398). June and Alex's desire to connect with Anangu culture through a sense-based experience on Country evokes Knudsen and Waade's (2010, 5) definition of "authentication" as a type of performative "production of authenticity … [as] a strategy to appropriate sites/places and … to invest emotionally in places". Rather than defining authenticity in relation to somewhat problematic claims about the objective qualities of things, people or places, it is "something experienced through the body", as a feeling in relation to a place (Knudsen and Waade 2010, 5). Knudsen and Waade (2010, 9) argue in favour of "re-investing in authenticity" as a concept, recognising that "performative authenticity not only signifies that we do and perform places by our actions and behaviours, but that places are something we authenticate through our emotional/affective/sensuous relatedness to them" (Knudsen and Waade 2010, 12–13). This form of authentication through direct embodied engagement is typically what visitors are seeking when they travel to places associated with Aboriginal cultural heritage to experience them in person, often by walking through Country, sometimes accompanied by a Traditional Custodian as a guide.

Consequently, "authenticity can be considered as a relational quality attributed to something out of an encounter" (Knudsen and Waade 2010, 13). This is in line with Wang's (1999, 351) understanding of "existential experience" as not only involving personal but also "intersubjective feelings activated by the liminal process of tourist activities". Wang (1999, 359) uses Daniel's (1996) account of tourists performing the rumba to explain that as they change from spectators into participants, and the rumba becomes an activity they perform, "it constitutes an alternative source of authenticity (i.e. existential authenticity)". The fact that the guided garden tour was an embodied experience in which June and Alex walked through the hotel gardens with an Anangu guide and had the opportunity to touch and smell the plants, is an important dimension of existential authenticity for some visitors. Similarly, my artistic activity of plein air painting was an embodied practice that I engaged in during my Uluru fieldwork, which allowed me to connect with and authenticate the place (Figure 5.1). The idea of existential

Figure 5.1 Evening, "Rangerville", UKTNP. Photograph by the author of a painting by the author.

authenticity, when accompanied by ideas of embodiment and "doing" place, or conceiving place as something we authenticate through our interactions with it (Rickly-Boyd 2013), may thus contribute to our understanding of what makes Aboriginal-led cultural tours memorable and perhaps transformative for visitors.

Conclusion

I argue in this chapter that visitor engagements with Aboriginal cultural heritage in protected areas were influenced by discourses about Aboriginal peoples that circulate in the settler-colonial Australian context. Some visitors attributed intractable deficits to Aboriginal peoples which they believed they had responsibility for remedying, arising from an adoption of whiteness as a universal norm (Erickson 2018). Other visitors assumed that there was such a thing as an "authentic" Aboriginality, characterised by an ongoing connection to an ancient Aboriginal culture, maintenance of language and living in harmony with a pre-existing nature, a form of "primordialist authenticity" critiqued by Theodossopoulos (2013b, 398–9). In contrast, "inauthentic" Aboriginal peoples were associated with a deficit discourse in which their urbanisation led to impurity and danger, and to their identities as Aboriginal peoples being questioned (Sissons 2005). There were parallels between visitor discourses about Aboriginal peoples and landscapes, since places on the urban fringe were also perceived as degraded, while Central Australian landscapes were viewed as spiritual and unspoiled. This is despite the fact that they have also been significantly altered by colonisation. Such discourses dehumanise Aboriginal peoples, by reducing them to an aspect of landscape and perpetuating a nature/culture dualism. Furthermore, they point to the way in which tourism is enmeshed with ideas about Aboriginality that circulate within the postcolonial Australian context and that have exerted power over Aboriginal and Torres Strait Islander peoples in material ways, both historically and contemporaneously (Dalley 2020; Klein 2020; Leckie 2021).

However, visitor experiences and engagements with Aboriginal cultural heritage in protected areas cannot simply be dismissed as reinscribing damaging settler-colonial discourses. Visitor motivations were sometimes multi-faceted and complex and could involve a genuine desire to come into relationship with Aboriginal peoples and places, through direct engagement and unanticipated embodied encounters (Knudsen and Waade 2010). In this sense, the type of authenticity visitors were seeking involved relationality, in that it entailed interactions between human and non-human others in place, interactions that triggered emotions and affects. One aspect of this desire for genuine engagement was the strong preference of some visitors for Aboriginal cultural tourism to be provided directly by the Traditional Custodians of the place being visited, or at least designed with their input. There was a clear expectation on the part of some visitors that such products, whether these be interpretive signage or digital forms of interpretation, should feature Traditional Custodians' voices and perspectives. By insisting on Aboriginal involvement and control of representation of self and culture, these visitors were in fact asserting the importance of Aboriginal sovereignty within a cultural tourism context.

In the next chapter, I explore the significance of embodiment, affect and emotion to visitor encounters with Aboriginal cultural heritage in place. Visitors' affective and emotional responses were a vital aspect of their experiences as they engaged with other visitors, the place, its history and their Aboriginal guides, suggesting that discursive factors alone are inadequate to explain these encounters. However, as will be explored, affective responses of pride, guilt and anxiety, which can be understood as "settler structures of feeling" (Rifkin 2011, 350), are also a product of discourse and can serve to reproduce the social, political and cultural dynamics of settler-colonialism.

Notes

1 The Parks Australia staff member who offered the ranger guided walk provided an anecdote about Aboriginal peoples using their newly built houses for firewood to illustrate what he understood to be a significant cultural gap, historically speaking, between traditional Aboriginal communities and Western culture. Several visitors I later interviewed picked up on this comment in their narratives.

2 Advertising by Voyages Ayers Rock Resort, Sydney Morning Herald Good Weekend Magazine, 8 June 2019.

6 Affectivities

"Settler structures of feeling" in colonised landscapes

Introduction

In this chapter, I explore visitors' affective engagements with the diverse forms of interpretation experienced at the various research locales, including Traditional Custodian led cultural tours, place-based interpretive signage, mobile phone audio tours and more museum-like environments such as the Uluṟu-Kata Tjuṯa Cultural Centre. I compare responses to different interpretive content and media, including interpretation developed and delivered by Aboriginal peoples or in collaboration with them, to interpretation developed in a non-collaborative way. I also analyse the affective, emotive and embodied engagements of visitors with the heritage landscapes they encountered, as a constitutive element of their experience of being in place. In so doing I interrogate how "emotions, sensations, and psychic life take part in the (ongoing) process" through which settlers continue to exert "authority over Indigenous peoples, governance, and territoriality", including in everyday ways that "help naturalize non-Natives' sense of entitlement to (once) Indigenous lands …" (Rifkin 2011, 350).

I foreground the framing of affect and emotion in the emerging field of settler-colonial studies because it has particular synergy with the visitor narratives emerging from my research (Mackey 2014, 2016; Rifkin 2011, 2013; Slater 2019). Mark Rifkin (2011) coined the term "settler structures of feeling" after the literary critic Raymond William's use of the term "structures of feeling" to describe "the ways in which social structures are registered as forms of individual and collective affect that do *work* in continually reconstituting social life" (emphasis in original, Williams 1977, 128, cited in Rifkin 2011, 361). Writing in the North American context, Rifkin explains that "settler structures of feeling" inform settler subjectivity and individual and collective identities, including how they understand their relationality with Indigenous peoples whose land their forbears claimed, and with this land itself. This is a reciprocal process through which "… institutional frameworks of settlement … give rise to certain modes of feeling, and … particular affective formations among non-Natives normalize settler presence, privilege, and power" (Rifkin 2011, 350). It is the affective modes of being in a tourism

DOI: 10.4324/9781032654973-6

context that "normalize settler presence, privilege and power" on which I focus in this chapter.

Debates around "settler structures of feeling", affect and emotion in a heritage context

The term "settler structures of feeling" implies that the feeling states that arise among non-Indigenous people under settler-colonialism can have certain typical characteristics that both emerge from and reinforce the system of settler-colonialism itself. Building on Rifkin's work, Mackey (2014, 2016) argues that in the Canadian context such "structures of feeling" are closely related to settler desires to lay unassailable claim to Indigenous lands, prompting anger at land claims by Indigenous Canadians. She links this anger to the settler belief that national belonging should cancel Indigenous claims to separateness. However, Slater (2019), writing in the Australian context, reveals how various feeling states that express a concern for the welfare of Aboriginal peoples are no less "settler structures of feeling" than the anger towards Indigenous Canadians identified by Mackey. As Slater (2019, 8) argues, "[w]hite anxiety – configured here as guilt, worry and pity – about the legacies of colonialism", also "works to neutralise the politics of sovereignty, and renew settler certainty and ethical belonging". Using the idea of "settler structures of feeling" as a conceptual framework, the range of affective states experienced by visitors who took part in my research, such as shock and surprise, anxiety, alienation, anger, pride, shame and remorse, can thus be understood as arising in the context of "[p]rocesses and institutional frameworks of settlement" (Rifkin 2011, 350). This includes emotional and affective responses in a more positive register such as awe and wonderment in the face of the time-depth of Aboriginal cultures and respect for their "sustainability".

Although the idea of "settler structures of feeling" serves to contextualise the affective and emotional responses of visitors, I also utilise understandings of affect arising from a Spinozian vitalism, as found in the work of Gilles Deleuze and Félix Guattari (1987). As Cameron Duff (2010, 882) explains, in these terms,

> … affect ought to be understood as a specific manifestation of a body's 'power of acting', its lived force or action-potential (Deleuze, 1988, page 50), for every affect is experienced both as a particular feeling state and as a distinctive variation in one's willingness or capacity to act in response to that state.

Affect is generally understood as an "… intense, embodied, physiological response …" that may be preconscious and transpersonal and to be linked with the concept of embodiment, which is conceived as "experiential, senses-inclusive meaning making" (Sather-Wagstaff 2017, 12). However, as Slater (2019, 13) argues, referring to Berlant (2011, 52), affect is more than "unmediated relatedness", since, as Berlant contends, the "visceral response is a trained thing, not just autonomic activity", mediated as it is by "memory and the past". Although there is much debate on the extent to which affect is involuntary or mediated by discourse, in the context of my research it seems that affect is bound up with the discursive in ways it is difficult to disentangle (Hoare 2020; Smith and Campbell 2015; Wetherell 2013; Wetherell et al. 2018). The discursive informs the affective, but affects are also co-constituted by assemblages of other "objects" coming together in space and time and are far from predetermined or predictable. The concept of affect thus understood is particularly useful for research such as mine, which deals with the way visitors respond to heritage interpretation in place, as it allows for the production of meaning by visitors in relationality with other visitors, Traditional Custodians and the more-than-human, in space and time. Although I recognise the centrality of settler-colonialism in informing visitors' felt responses, the concept of affect accounts for the way that meaning is also created in the act of doing.

Furthermore, emotion and affect can prompt transformation at the individual and broader societal level, since feelings that inspire or disturb can provoke critical reflection upon the self and the world to achieve social justice goals (Witcomb 2013, 2015a, b). Andrea Witcomb (2013) argues that poetic forms of curatorship and interpretation in museum and heritage contexts that prompt affective and emotional responses that unsettle are much more effective than didactic approaches in engaging and "transforming" visitors. Other scholars agree that embodied, sense-based engagement can stimulate affect at heritage places, and this, in turn, can produce empathy, an emotional and cognitive capacity linked with such transformations (Gokcigdem 2016; Gregory and Witcomb 2007; McRainey and Russick 2010). Affect and emotion are therefore not just manifestations of "settler structures of feeling" but can work to undermine and transform such structures through empathic engagements, including those that enable a type of productive shame that Sara Ahmed (2004, 120) refers to as "shame that is not only before others, but for others".

The importance of affect, emotion and embodiment is increasingly recognised in fields such as critical heritage, museum and tourism studies, which deal with the way that people respond to heritage in place and visitor engagement in museums. Divya Tolia-Kelly et al. (2017, 1) note a shift in heritage scholarship, in line with approaches in cultural geography, to "more-than-textual embodied approaches [that] ... engage with experience, the sensory realm and the affective materialities and atmospheres of heritage landscapes". As David Crouch (2015, 178) succinctly explains, considering "the phenomenological and performative character of living, ... encounters with heritage sites and events *happen*" (emphasis added). Heritage itself is increasingly understood as a verb rather than a noun, a performance that visitors enter into in the context of site interpretation that creates meaning and value, and not a fixed or static "thing" (Harrison 2013; Harvey 2001; Smith 2006; Waterton 2014). For this reason, Waterton (2014, 830) argues that the concept of affect is useful for understanding the way that people interact with heritage, particularly places of difficult heritage that evoke strong reactions, but also more general embodied responses to "multi-sensual sights". Contested landscapes such as Uluru-Kata Tjuta National Park constitute difficult heritage of this type, although my research suggests that any place that includes the material cultural heritage of Aboriginal peoples could present challenges to visitors in a settler-colonial context.

Emotion, too, which unlike affect can be understood as a conscious feeling, although commonly regarded as subjective, is increasingly understood as "relational flows, fluxes or currents, in-between people and places" (Bondi et al. 2007, 3). There are commonalities between heritage studies and a geography of emotions that seeks to explore the emotional connections that people have with place, and the other human and non-human beings who inhabit it (Davidson and Milligan 2004; Kearney 2009; Pile 2010; Thien 2005). A parallel turn in landscape studies to viewing landscape and the body as co-constituting in an ongoing process of becoming is thus to some extent mirrored within critical heritage studies in the recognition that heritage is both a process and performance rather than a pre-existing materiality (Duff 2010; MacPherson 2010, 2016; Metro-Roland and Soica 2019; Smith 2006; Waterton 2014, 2019). These understandings share commonalities with Indigenous onto-epistemologies that conceive the landscape as agentic, and thus as a changing, active entity co-creating human experiences and not as an inert surface on which living occurs (Tuck and McKenzie 2015).

The importance of affect and emotion in visitors' engagement with museums and heritage sites has been highlighted in recent research by Laurajane Smith (2014, 2021). In fact, she reveals that some people purposively visit these places to emotionally engage with them. As a consequence, she observes that "the learning/recreational paradigms of understanding visiting is far too limited" (Smith 2014, 125), concluding that embodiment, affect and emotion are central to the way heritage, in all its forms, is created and imbued with meaning. Related research within the field of critical heritage studies further demonstrates that the affirmation of emotions, identity and beliefs are key motives for heritage and museum visitation, and embodied emotional cognition and affective engagement are important aspects of performances at heritage places (Dittmer and Waterton 2019; Micieli-Voutsinas 2017; Smith and Campbell 2015; Tolia-Kelly et al. 2017; Waterton 2014; Waterton and Watson 2015).

Smith and Campbell explain that visitors to museums and heritage places experience a range of emotions, which they refer to as "registers of engagement" (2015, 444). Smith (2014, 126) argues that "emotionally flat responses may … close down critical engagement and reinforce understandings of and commitments to consensus views of history". In contrast, she contends that empathy is important in "the development of critical and progressive insights into the past and its meanings for the present" (Smith 2014, 126). Smith and Campbell (2015, 450) observe that there is now a body of research in the museum and heritage field that demonstrates how a deep empathic response can lead to "greater engagement with hidden or marginalized histories" (see also Bozoğlu, 2020; Gokcigdem 2016; Witcomb 2013, 2015a, b). This is of particular relevance to visitors' engagement with places of cultural significance to Aboriginal peoples, as Aboriginal peoples have historically been marginalised by the settler-colonial state. As a consequence, histories of their relationships with places such as my research locales have often been hidden or unavailable. Moreover, as Witcomb (2013) argues, and as is taken up in the next chapter, affective and empathic engagement at museums and heritage sites can trigger processes of personal transformation, where one comes to question what one thinks and believes.

As Amy Coplan (2011, 5) explains, "empathy is a complex imaginative process in which an observer simulates another person's situated psychological states [both cognitive and affective] while maintaining clear self-other differentiation". It thus goes beyond the more common understanding of empathy as a sympathetic emotional connection to another (Feshbach and Feshbach 2009). Empathy is both cognitive and affective and these aspects are interdependent, since empathy is at the same time "a process of understanding and emotionally responding to the thoughts and feelings of others" (Savenije and de Bruijn 2017, 833). While moral philosopher Michael Slote (2007, 8) regards empathy as "essential to caring moral motivation", there are different levels of empathy and the varied ways in which it can operate have implications for its capacity to effect personal and political transformations. Smith and Campbell (2015, 450) identify "deep empathy", which "moves beyond the 'I feel sad' for this other", as "key to triggering a response that engages the imagination in such a way that visitors start to question what they know and understand". Similarly, Coplan and Goldie (2011, xxxiii-xxxiv) explain that "higher-level empathy", also known as "perspective shifting", requires imaginative investment and "perspective-taking ... [which is] other-oriented rather than self-oriented".

In contrast, lower-level empathy, which "simply mirrors another's emotions" can lead to an "overidentification with victims of past [or present] injustice", and this, in turn, can result in "the appropriation or colonisation of their identities" (de Simine 2013, 46). In a settler-colonial context, empathy can also arise from an "impulse ... to overcome difference" (Jones and Jenkins 2014, 12), and to deny the cultural distinctiveness and embodied socio-cultural differences of Aboriginal peoples. As demonstrated by Edith's response to the Uluṟu Culture Centre, discussed below, empathy can also manifest as what Nilsen and Bader (2016, 118) refer to as "egocentric bias that leads us to assume that others think the same way that we do". This may be one reason why, as Coplan and Goldie (2011) explain, empathy is not always welcomed by its object, and can be experienced as intrusive and inappropriate. Empathy can also resemble or engage the type of concerned anxiety and pity that Slater (2019) identifies as a settler structure of feeling, framing Aboriginal peoples as constantly in need of white help and intervention, thereby continuing to affirm white authority.

Although I recognise the problematics of empathy in a settler-colonial context, my research suggests that such responses can be important to the capacity of visitors to drastically change their perspectives on culturally insensitive activities they were invested in, such as the Uluṟu climb. In contrast with the transformative power of deep, empathic engagement, Smith (2014, 127) concludes that for the vast majority of museum and heritage visitors interviewed in her research, "the visit is about reinforcing not only what they already know, but more importantly, what they already feel and believe". While my research suggests that this is certainly the case for some visitors, particularly those who travelled to places such as Uluṟu-Kata Tjuṯa National Park to reinforce national narratives of pride and conquest by climbing Uluṟu (Barnes 2011; Hueneke and Baker 2009; James 2007; Waitt et al. 2007), other visitors appeared changed by their experiences in ways they could not have anticipated.

Despite the importance of such changes, "settler structures of feeling" arising in visitor narratives were both persistent and pervasive. Visitor responses were thus not only constituted in what Lorimer (2005, 84) calls "shared experiences ... fleeting encounters [and], embodied movements ...", but also arose from deep seated feelings of shame and guilt associated with beliefs about the permanent "damage" the settler-colonial state has inflicted upon Aboriginal peoples. Encountering a place such as Uluṟu-Kata Tjuṯa National Park, so commonly perceived as an icon of Aboriginal Australia, seemed to trigger such responses for some white Australians, bringing to "top of mind" a sense of responsibility for past injustices and the present perceived "disadvantage" of Aboriginal peoples. Although, as Tolia-Kelly (2016, 901) asserts, "[a]ffective atmospheres are everywhere and coalesce in the spaces of heritage", these atmospheres were particularly intense in Uluṟu-Kata Tjuṯa National Park. As visitors' affective responses to interpretation within Uluṟu-Kata Tjuṯa National Park were complex, shared and multi-layered, this chapter predominantly focuses on responses in this locale, while also integrating relevant visitor narratives from Red Hands Cave and the Jenolan Caves, including the particular perspectives offered by Aboriginal visitors.

Empathic engagements in Uluṟu-Kata Tjuṯa National Park

Visitors interviewed within Uluṟu-Kata Tjuṯa National Park strongly approved of the interpretive signage located on walks in proximity to Uluṟu, believing that it deepened their capacity to understand the landscape and the importance of this place to Aṉangu (see Chapter 5). This signage built on the existing resonances within the landscape arising from decades of visitation and the narratives already attached to this place (Tolia-Kelly 2016), although it also served to reframe these. It was hardly surprising, then, that visitors used affective language to describe their impressions of this interpretation as "touching", "powerful" and "grounding", suggesting that it prompted an immediate, physical connection with the place. They reported that experiencing the signage in-place increased their understanding, provided insight and fostered empathy towards A_n_angu, through the affective engagement it invited.

The signage formed one part of a knowledge producing assemblage also made up of their own moving bodies and those of other visitors, the landscape and other aspects of the environment such as the weather. The type of empathy experienced by visitors in response to the on-site signage assemblage could lead to an opening up of interest based on growing regard for A_n_angu, initiate shifts in perception, and, less commonly, prompt deep, potentially transformative insights. The capacity of the interpretive signage to draw visitors in may also have contributed to its success as a form of pedagogy (Witcomb 2013), as visitors described it as interesting and informative, and had good recall of its content. Through its capacity to affectively engage visitors, Ulu_r_u-Kata Tju_t_a National Park's interpretive signage successfully fulfilled a pedagogical function but was not limited to this.

The presentation of the content of the on-site signage in a collective A_n_angu voice, and the orientation of the interpretation to subjects important to A_n_angu, such as the three major creation stories that form part of Tjukurpa, also facilitated a strong level of visitor engagement. Additionally, visitors appreciated the use of the Pitjantjatjara language in the signage, perhaps because it emphasised the voice and presence of A_n_angu. As outlined in Chapter 3, much of this signage also spoke about A_n_angu culture in the present tense as a living continuing culture. However, for some visitors there was a sense of incongruity between the powerful presence of the A_n_angu voice within the interpretive signage and what they perceived as the physical absence of A_n_angu within the park, prompting unanswered questions about the way A_n_angu currently engage with this landscape and with the tourists who visit it. The "absent presence" of A_n_angu unsettled some visitors (Hill 2015), prompting them to speculate that tourism may have inhibited the use of the park by A_n_angu and leading to a level of anxiety about their role as tourists (Slater 2019), as will be discussed below.

An awareness that settler-colonialism was a pervasive but hidden presence within the park came through in Fran's discussion of the Mala Walk at the base of Ulu_r_u as a better way to form a relationship with this place than activities such as climbing Ulu_r_u. As Fran explained, the interpretation provided by the on-site signage on the Mala Walk,

> … brings in the stories that make this place so significant, it is not just the rock and you climb up it, and if you don't climb up it, you walk around it … Like as far as there being another way to connect, and learning some of the stories … Um yeah, I've enjoyed it [the Mala walk] for that part actually …

This signage on the Mala walk thus provided Fran, introduced in Chapter 5, with "another way to connect", the use of the word "connect" invoking the intimate sense of touch, an act I witnessed in the park when visitors

Kuniya walk and Mutitjulu waterhole, May 2019

I encountered two young European tourists at the Mutitjulu waterhole. The male jumped down onto the sand, looked around at me as I walked up, said "Shhh!", and smiled conspiratorially. There's no signage there saying not to, but he knows he shouldn't cross the roped off barrier to walk on the sand.

Further on, are a family with a young child. There is a sign here that says "no climbing" as the gradient of Uluṟu at this point makes climbing possible. Glancing at me, the father instructs his son, "You can't climb but it's OK to touch Uluṟu". The child's mother affirms, "You can say you touched Uluṟu".

reached out to touch Uluṟu or spoke about the desire to do so. On occasions, these engagements transgressed the boundaries set for them, as described in the text-box opposite. These moments of witnessing were made more awkward by visitors attempting to elicit my support for their behaviour, on the basis that I was a visitor like them. But rather than expressing solidarity with them, I wanted to distance myself from those visitors who did not respect the boundaries set for them within the park. My affective response was more aligned with those visitors I interviewed who complained about seeing footprints in the sand in roped off areas, identifying it as a mark of disrespect for Aṉangu and the national park itself.

Undertaking short walks within Uluṟu-Kata Tjuṯa National Park, particularly at the base of Uluṟu, provides one of the only means for some visitors to go beyond a surface-level response such as simple aesthetic appreciation and come to understand meanings of Country to Aṉangu. The majority of visitors stay in the park for only one to three days, and self-drive visitors, rather than those on guided tours, only have access to the interpretation provided in on-site signage, particularly along the popular walks at the base of Uluṟu, and at the Uluṟu Cultural Centre. Despite the limitations of this medium, reading this signage and learning about the meanings of place to Aṉangu aided some visitors in forming a sense-based bond with the landscape, as invited by the panels at Kantju Gorge.

This signage exhorted visitors to "feel Country" by listening to the sounds around them and drawing Country in through the breath (see Chapter 3). The comments about this signage by Anne and Mark, two young visitors from Victoria, reinforce Smith's (2014) argument that educational and recreational paradigms are unable to fully explain the reason that visitors travel to heritage places and engage with heritage interpretation. Mark, for example, explained that the signage encountered on the Mala Walk, "grounded me and made me feel more present and pay attention to the silence, and to the request to keep the noise down and made me tune in a little bit". Mark continued "[t]o listen to the rustling of the trees .... We'd also noticed before that they're bits where the rock whistles in the wind, so we'd already been looking out for the sounds". While they were already attentive to Country, following the instructions of the interpretive sign (Figure 3.4) to be quiet and "sit and think about what you are experiencing", deepened their connection by encouraging a meditative yet embodied response that engaged other senses beyond that of sight (Knudsen and Waade 2010). As visitors like Howard also articulated in the previous chapter, it is this feeling of connection occurring in space and time that visitors are seeking when they travel to heritage places and engage with interpretative signage, rather than an experience that is purely educational.

Background noise, UKTNP, May 2019

A very busy day. Felt disappointed I didn't get back in time to experience the end of the day on the back veranda of the house I am living at in the park while conducting my fieldwork. It nestles behind a sand dune, close to Uluṟu. The quiet, the fading light and the crickets have all become important to me.

But this evening there's some type of machine noise like an air-conditioner motor. I've not noticed it before. I can even hear it inside. When you're used to perfect silence, a noise like this can become really noticeable and disturbing. I can't stop hearing it, move my attention away from it.

If I was at home, it would probably be just another form of background noise, blending in with its surrounds.

Like Mark, I also valued quiet times in the park and found that I became very attuned to the sounds around me as a way to connect with the spirit of place, as I reflect in the text box above. Similarly, Naomi, introduced in Chapter 3, had a strong affective reaction to the Aṉangu collective voice on signage at the climb site,[1] stating, "I thought it was really powerful ... When they're saying, 'I asked you not to [climb Uluṟu]', and I'm thinking [big indrawn breath], if that doesn't *hit you in the heart*, what does? Really?" (emphasis added). The capacity of interpretation within Uluṟu-Kata Tjuṯa National Park to "hit you in the heart", a phrase that conveys an almost physical affective impact, was not confined to place-based signage.

I interviewed a number of visitors immediately after they had walked through the Tjukurpa tunnel and watched the "hand-back" video in the park's Cultural Centre. The video tells the story of the "hand-back" of the park to Aṉangu from an Aṉangu perspective, including the years of negotiation that Aṉangu engaged in with the Australian Government and the vociferous opposition they encountered from the NT Government. The affective power of the video is conveyed by one visitor's choice of the words "incredibly moved" and "touched" to describe his reaction to it, terms that evoke both changed states of feeling and the tactile sensation of physical touch. A number of other visitors responded to the video in both empathic and emotional ways, able to articulate altered states of feeling such as an increased respect for A<u>n</u>angu and their land rights' struggle after viewing it, suggesting that the video had a role in visitor transformations (Witcomb 2013).

For example, Eric, introduced in Chapter 5, noted that the video dealt with,

> ... one of the A<u>n</u>angu leaders, when they were trying to organise that [hand-back], all of the flights they had to take down to Canberra ... I suppose the thing that struck me about that was how alien that must have been to the A<u>n</u>angu people, and again they were the ones having to go out of their comfort zone, going out of their environment to ultimately fight for something that was previously theirs. Ehm [laughs], so I thought they weren't exactly dealt a very fair hand ... but I have a lot of respect for the fact that they continually went out of their comfort zone to fight for that.

Eric's account of his response to the video demonstrates the "complex imaginative process" that characterises empathy (Coplan 2011, 5). Eric employs this process when he considers how alien travel to Canberra must have been for A<u>n</u>angu. This resulted in the "perspective shifting" of increased respect for A<u>n</u>angu and their fight to reclaim their land that he describes (Coplan and Goldie 2011, xxxiii). Both his imaginative investment and perspective-shifting signal that he was engaging in what Coplan and Goldie (2011, xxxiii) refer to as "higher-level empathy" rather than simply mirroring the emotions he observed in the video. His empathic reaction served to deepen his engagement with the marginalised history of Anangu's land rights struggle (Smith and Campbell 2015). Eric's response suggests that visitors identified with, remembered and were changed by interpretation with an affective and emotional dimension such as the "hand-back" video, in which A<u>n</u>angu quite openly discussed their joy at hand-back and their determination to never give up their land (Witcomb 2013). The video also dealt with contemporary post-contact history and political themes, and visitors' responses suggest a willingness to engage with such themes in the context of protected area interpretation.

Like Eric, Jennifer, a middle-aged domestic Australian visitor, also responded emotionally to the "hand-back" video, but rather than prompting a feeling of empathy, her response was one of pride. She explained her feelings as "proud that we did that. You know it felt good that we did the right thing. We did lots of wrong things, governments, and ... And it actually, it felt good". The word "we" appears at least three times in Jennifer's narrative, revealing that her focus was on the actions of white Australians such as herself, and perhaps also reflecting a desire to elicit my own participation in this narrative as an assumed fellow white Australian. Similarly, Carlos, a Chilean-Australian, when interviewed on the base walk around Uluṟu, summed up his response to the video by saying "I feel good, because they gave it back to them and they haven't taken it away and I thought well, that's good". While Carlos' sentiments echoed Jennifer's in the sense that the "hand-back" video prompted positive feelings, he did not include himself in the collective beneficent white "we", nor did he espouse feelings of pride in "hand-back". Unlike Anglo-Australian Jennifer, Carlos, as a first-generation migrant to Australia, did not include himself in collective responsibility for this action, and did not share in the same "settler structures of feeling" arising from it.

Jennifer, perhaps reluctant to feel difficult emotions, grasps at the feeling of pride "that we did the right thing" by handing back the park to Aṉangu, closing off the potential for deep transformative change that might be prompted by an emotion like shame, as discussed below in relation to Derrick (Probyn 2005). An emotion such as pride represents a feeling-based response that, as recognised by Smith and Campbell (2015), could undermine a social justice agenda by shutting down the potential for behaviour change and the broader social and political changes arising from it. The "hand-back" of the park to Aṉangu is a story of postcolonial Australia, so to speak, with a happy ending, signalling that nothing more needs to be done. As Ahmed (2010, 50) notes, characterising good feelings like happiness as open and generative and bad feelings like shame or anger as closed "allows historical forms of injustice to disappear". Accordingly, rather than leading to a radical change in her intention to climb Uluṟu, Jennifer's visit to the Cultural Centre merely prompted her to reconsider this decision.

However, for other visitors such as Derrick, the capacity to affectively and emotionally engage with forms of interpretation such as the "hand-back" video resulted in a dramatic change in thinking that stimulated a critique of his own racism. Prior to his arrival in the park, Derrick's determination to climb Uluṟu was, by his own admission, based on a discourse of rights, viewing his right to climb as superseding Anangu's objections to the act. He explained that he travelled to the park to see Uluṟu and climb it, "but after going through the [Cultural] Centre and reading and learning a bit, I wouldn't climb it anyway now". He elaborated,

> Yeah, I came here to climb the rock and have a look at the rock and everything about the rock. Now, it's 180 degrees. It's good to see the rock and it's wonderful. But to *learn* (big emphasis), the significance of it to them (Aṉangu), and to see that gee I know nothing, I'm 59 years old and it's been a real eye opener for me, it's completely changed me in many ways in a matter of hours. I think when I first came here, I was actually quite racist through ignorance and now I really want to learn and I sort of want to know things and I want to sort of get their side of the story.

Derrick's response to the interpretation at the Cultural Centre asking visitors not to climb was very affective, evident both in his tone of voice and that he was several times close to tears.[2] His response was also emotional and empathic, revealing a deep rather than superficial or banal "register of engagement" (Smith and Campbell 2015, 444). He expressed a range of emotions, from feeling "ashamed about the way we've treated [Aṉangu] people", to "incredibly moved" and "touched" by the "hand-back" video, to "robbed" that "we weren't taught a lot of these things earlier on". Derrick explained that very little about Aboriginal culture had been taught to him at school and that he had been given "a very, very one-sided view of history ... Yeah, it was an empty land and we came along and we made it into this great country".

Derrick's response suggests that the way some white Australians position themselves in relation to Australia's colonial history colours their cognitive, emotional and affective engagement with the interpretation of Aṉangu culture in places like Uluṟu-Kata Tjuṯa National Park. My interview with Derrick also revealed that visitors had the capacity to significantly change their views on subjects such as the climb of Uluṟu as a result of exposure to interpretation developed in consultation with Traditional Custodians that included a direct collective Aṉangu voice addressing the reader/viewer. This intervention by Aṉangu may represent an example of what Foucault (1980, 201) has described as the operation of "local relations of power" that is to some degree independent of, and able to challenge, centralised power through "a multiplication of discourse" (Foucault 1984b, 311). Such "multiplications" have the potential to disrupt a hegemonic settler-colonial discourse or "regime of truth" (Hall 2001, 76) about Uluṟu as a landscape feature Derrick had a right to climb in favour of a localised perspective arising from an Aṉangu onto-epistemology.

However, Foucault's theories of the operations of discourse cannot fully account for Derrick's response, given that they were developed historically prior to the theorisation of affect and emotion as "social and cultural practices" (Ahmed 2004, 9). I argue that Derrick's capacity to feel higher-level empathy for Aṉangu, feelings triggered by watching the "hand-back" video and reading the text in the Tjukurpa tunnel about why Aṉangu ask visitors not to climb, was instrumental in facilitating the transformation of outlook he describes. Encountering these interpretive messages created a dilemma that generated affective responses such as pressure and anxiety (Mezirow 2000). In Derrick's case, this tension arose from the clear disjuncture between Aṉangu's request that he not climb Uluṟu and his stated intention to climb as the purpose of his long journey. Without the inner conflict this dilemma triggered, and the uncomfortable feelings it engendered, the epiphany Derrick describes might not have occurred.

However, as discussed further in Chapter 7, the extent to which a single experience, no matter how powerful, can be regarded as genuinely transformational of an ingrained mode of feeling, thinking and behaviour such as racism, as Derick claims, must be queried. As Smith and Campbell (2015) note, feeling-based responses are insufficient for achieving social justice, and may even undermine it, if they are not followed through with action, both in the short and longer term. Derrick's empathetic response is consistent with this social justice agenda, as it led to the decision *not* to take an action that would have been culturally offensive. The engagement of affect and emotion at museums and heritage places can thus lead visitors to shift their views on social justice issues in ways that translate into changes in behaviour, at least in the short term.

The politics of shame, white possession and "virtuous anxiety"

The feeling of shame, affective because it is experienced by the body, while deeply unpleasant, can also be instrumental in moving one towards the type of transformation Derrick describes. Shame, as Ahmed explains, is an "… intense and painful sensation that is bound up with how the self feels about itself" and is "usually experienced before another" (Ahmed 2004, 103). Probyn (2005) argues that shame can be a productive emotion because it

> … compels an involuntary and immediate reassessment of ourselves: Why am I ashamed? … Can I rectify the actions that have … brought shame upon myself …? Shame in this way is positive in its self-evaluative role; it can even be self-transforming" (Probyn 2005, xii).

However, shame, Probyn (2005, xii-xiii) further contends, can only be positive "when it is acknowledged", moving from the realm of the affective to the cognitive. Waitt et al. (2007, 253) go further, arguing that because shame "involves one's care for a situation", it can function as a "moral gateway" (Waitt et al. 2007, 258) on the road to meaningful reconciliation between Aboriginal and non-Aboriginal Australians.

Other scholars, by contrast, highlight the risks associated with shame, particularly when it is enacted on behalf of the nation, and such acts signal the closure, rather than the opening, of a conversation about reconciliation (Ahmed 2004, 2005; Kizuk 2020). Influenced by Ahmed's work on shame, particularly her discussion of the white subject feeling "bad" about its racism in order to feel "good" again, Kizuk (2020, 166) argues that there is a tendency for settlers to simply confess their shame so that they can feel good "for having completed introspective work". While Kizuk (2020, 168) acknowledges that there may be "forms of shame that have transformative value", she does not speculate about the pattern these may take. Ahmed (2004, 119) is more helpful, suggesting that in order to be useful, shame should not simply be performative, because "the act of speaking our shame does not undo the shame of which we speak". Recognising this would refuse the "conversion of shame to pride, in an act of shame that is not only before others, but for others" (Ahmed 2004, 120). It may be that Derrick's feelings of shame

Crime wave, UKTNP, May, 2019

Unsure what to expect, I take the turn off the ring road towards Mutitjulu in the early evening, to arrive at what will be my home for the next month in "Rangerville". The most prominent building I pass is the police station.

The ranger houses on the outskirts of Mutitjulu are surrounded by cyclone fencing and gates with padlocks. I was advised to drive my hire car inside because kids from the community might get to it. As soon as I saw it I knew that the ridiculous spoiler and mag wheels were asking for trouble.

As I fumbled with my gate lock on the first evening, an Aṉangu family passed in the street, the little girl yelling and waving "Hi! Hi!", "Hi", I yelled and waved back, bolting the padlock, feeling ashamed.

The following day, on my way out, I noticed two large police fourwheel drives parked at the front of the police station.

For a small remote town, this was a very visible police presence. Not many community facilities and a pool that closes in autumn/winter, but an unmissable police station, open all year round.

contributed to the transformational change he describes, towards not only the climb but the desire to learn more about Aṉangu and further transform his attitudes and knowledge by "getting their side of the story". This contrasts with a settler response that attempts to deny collective responsibility for past and ongoing injustices towards Aboriginal peoples through what Sarah Maddison (2012, 2) describes as "a rationalisation of colonialism as just, inevitable and ultimately for the good of a 'primitive' race".

Thus, Edith, an Anglo-Australian visitor in her 80s, interviewed at the Uluṟu Cultural Centre, described the access that Aṉangu provide to Uluṟu-Kata Tjuṯa National Park as an obligation, asserting,

> Yeah, but I really feel they've got to [welcome visitors to the park]. Like, they have obtained a lot from the white people like, as they say, the people who came over and took their land. I think they have gained a lot too, because they wouldn't be able to do a lot that they're doing now. Can you understand what I mean?

In a performance of "settler common-sense", Edith treats settler access to territories such as Uluṟu-Kata Tjuṯa National Park as a given, demonstrating how, as Rifkin (2013, 324) has argued, "the claiming of a naturalized right to Indigenous place lies at the heart of non-Native modes of governance, association, and identity". Stopping well short of experiencing uncomfortable emotions like shame or guilt, Edith relies on the rationalisation that colonisation has benefited Aboriginal peoples to justify her presence in the park.

There were parallels between the feelings of shame experienced by visitors and my own unanticipated moments of shame during my time in the park, captured in the text-box. Although I did not know I would be staying in a property enclosed by cyclone fencing when I was offered accommodation in the park, this fencing nevertheless set me apart from A̲nangu living in Mutitjulu, implicating me in the racial divide that existed in the park. I was starkly confronted with this state of affairs when an Aboriginal child greeted me while I was locking the gate to the fence enclosing the property I was staying in, prompting the spontaneous feeling of shame I describe, a form of shame that, as Waitt et al. (2007) argue, demonstrated my care for this situation.

Like shame, collective guilt has been described by social psychologist Sonia Roccas et al. (2004, 145) and colleagues as "never comfortable" and "especially painful" when related to situations of ongoing group conflict. However, unlike shame, which tends to arise from one's care for another, guilt is internally focused, and less likely to engender productive outcomes. In fact, Maddison describes guilt as a "profound and complex barrier to a national examination of the systemic racism that continues to marginalise Indigenous peoples in Australia and other postcolonial states" (Maddison 2012, 3). Guilt was a much more frequently mentioned emotion than shame in my visitor interviews, the two troubling feelings arising primarily in the narratives of retired domestic Australian visitors to Ulu̲ru-Kata Tju̲ta National Park. The fact that similar feelings were not expressed by visitors to the Blue Mountains National Park or the Jenolan Caves suggests that Ulu̲ru-Kata Tju̲ta National Park held meanings to visitors that the other two locales did not. As a remote location in Central Australia, Ulu̲ru was for white Australians highly symbolic of Aboriginality, prompting feelings about Australia's colonial history that the whiter and more urbanised Blue Mountains and Jenolan Caves did not.

Jane and Arthur, a couple in their 80s, participated in the Patji cultural tour with an A̲nangu host on APY lands outside of Ulu̲ru-Kata Tju̲ta National Park and later took part in an interview with me at their Yulara hotel about this experience. Although they remained silent when our A̲nangu host spoke to the group about his family Tjukurpa, Jane and Arthur later revealed that these accounts had a significant emotional impact on them. Jane responded with the clear moral judgement, "these things [such as child removals] should never have happened." She also explained that when she heard the A̲nangu host's account of his family's experience of the Stolen Generations, she had "wondered what it would actually do to a family." Jane thus demonstrated a degree of empathic engagement with Aboriginal families who experienced the devastating impact of child removal under Australian Government policies of protection and assimilation (Armitage 1995). Her final comment, "[a]nd still we haven't made amends", highlights that in her view settler Australians have not adequately attempted to address past wrongs, as if any form of simple atonement were possible (Maddison 2012). Jane's voice trembled on these words, an affective reaction suggesting strong underlying emotion.

While the desire to atone for a wrongdoing is an understandable human emotion in this context, there may be a degree to which this eagerness for solutions is also a desire for closure, to distance oneself from uncomfortable emotions and to assuage unpleasant feelings of settler guilt (Ahmed 2004). Both Jane and Arthur identified feelings of guilt about post-contact Australian history, Arthur explaining "I carry with me a residual guilt, that I'm part of what has changed this continent". Arthur's words evoke this guilt as an abiding weight that he physically "carries" with him, constructing it as an enduring burden that is part of his identity as a settler Australian. Similar to Naomi and Ray, discussed below, this guilt was not just about the impact of settler-colonialism on Aboriginal peoples but also its perceived damage to the Australian landscape.

These narratives have much in common with what Slater (2019, 3) describes as "virtuous anxiety", a form of white "worrying about Indigenous people, which ... affords catharsis and satisfaction", but in so doing "... displaces the political". She explains, "[g]ood white people take responsibility for past injustices, face up to colonial history and the ongoing destruction and marginalisation, often by feeling guilty and pitying Indigenous people, or empathising. We work to make amends" (Slater 2019, 9). Virtuous anxiety is thus a deeply embedded feature of settler structures of feeling that is, often unconsciously, invested in reinforcing settler authority by constructing Aboriginal peoples as victims.

For example, Naomi and Ray articulated a feeling of uncertainty and an amorphous anxiety about the legitimacy of their own decision to visit the park, describing this as,

> ... the conflict of are we helping or hindering? Or should they have people here at all? ... we've just done so much *damage* to them culturally ... And I feel like we're just rubbing their faces in it and, yeah, are we any better [than climbers] by coming here and staying in fancy hotels? (Ray, emphasis added).

Ray explained that this "moral battle" led him to,

> feel, like, a *real guilt*. I feel like we've taken away from the animals as well, ... obviously hundreds, thousands of people coming to those places constantly, we've driven the animals away ... And so you know you want to see, experience these places and so does everyone else. You got to look at your impact (emphasis added).

There was in fact a drought in the region of Central Australia where the park is located in May 2019, when I spoke to Ray, which meant that larger animals had departed to find food and water elsewhere. Ray's guilt, then, also prompted a form of anthropocentrism, which explained the behaviour of animals primarily in terms of tourist impacts, when climatic conditions were also responsible. Ray and Naomi's dialogue about their decision-making around either purchasing an artwork from Aṉangu women in the Yulara town square, or from an art gallery at the Cultural Centre within the park, spoke further to these settler anxieties. Naomi explained,

> I think what we've found hard, too, is, like I said, I'd really like to buy a painting, you know, and the kids all had a little bit of money to spend and bought something at the Cultural Centre ... And we were saying, 'Is this the best way to support people buying it through a place like that?' I find that really hard as a tourist.

> *Ray:* If you buy direct from the women out in the town square, you don't know, are you helping them by giving them the money direct and that way they get all the money and they get to choose what they do with it, or are you adding to the problem? Are they gonna go and buy alcohol?
>
> *Naomi:* (interrupting) But who are we Ray, who are we to say how you spend your money...
>
> *Ray:* Someone like us, we don't know what commission the gallery takes and we don't know if it's frowned upon, like, if we're doing the right thing or are we impacting in a negative way or a positive way.

Like their concern about their overall impact as tourists within the park, Ray and Naomi's desire to purchase an artwork by Aṉangu became an ethical dilemma, one which revealed their care for the situation (Waitt et al. 2007) and their tendency to be reflective about their impact as visitors. However, it also revealed a paternalistic "virtuous anxiety" (Slater 2019, 3), particularly on the part of Ray, not to exacerbate the perceived problem of alcohol abuse by Aboriginal peoples by paying for an artwork in cash.

Sianne Ngai (2005, 7) describes anxiety, like irritation, as a minor but "ugly feeling" that is "defined by flatness and ongoingness" and, unlike fear, is not sudden or arising from a particular event or threat. While fear is likely to be resolved when the threat passes, because it "lacks a clearly defined object" (Ngai 2005, 21), anxiety persists but in an amorphous or ill-defined way. Although Ngai (2005, 22) notes that emotions are often implicated in actions, what is noteworthy about the "ugly feelings" she investigates is that they "tend to be diagnostic rather than strategic, and to be diagnostically concerned with states of *inaction* in particular", (emphasis in original). Anxiety thus resembles one of the banal emotions or affects Smith (2014) describes that have the potential to shut down rather than enliven visitor engagement with heritage places, all the more so in the case of "virtuous anxiety", because it facilitates a form of white catharsis.

Becoming home, Rangerville, UKTNP, May 2019

Green grass is sprouting incongruously in the red earth of my backyard after rain. It looks so wrong.

No wildflowers as yet. And no frogs singing-learnt they need a lot more rain. But some lovely gurgling at Mutitjulu waterhole. The seepage is visible on Uluṟu for days.

Reading this journal entry much later I realised I had not only lost track of the date, but had begun to call Rangerville "home" and to refer to the outside area of the property where I was staying as "my backyard", reflecting both my growing attachment to the place and a settler attitude of possession that is both unconscious and persistent.

Catharsis can also be achieved by taking symbolic steps to achieve atonement, as was the case with Naomi's act of travelling to Uluṟu to return sand that had been brought back for her when she was a teenager, and in the case of Carlos, for climbing Uluṟu and taking two pieces of rock with him 25 years prior. As forms of identity work, they resemble the affective practices of visitors in museums who engage with exhibitions to "construct, maintain, and adapt" (Rounds 2006, 133) their sense of self. The actions that Carlos had taken many years ago did not sit well with his current self-concept and required some form of symbolic reparation to restore his sense of self. Although cathartic, these actions were in themselves apolitical, assuaging settler guilt but doing nothing to achieve justice, reparation or restitution for Aboriginal peoples in the present. Naomi's and Carlos' narratives about taking and returning parts of the physical landscape of the park, which there is no scope to reproduce here, are also important from another perspective. They reveal visitors' affective desire to touch and claim parts of the landscape and are part of a broader phenomenon discussed by Jasmine Foxlee (2007) in her research on the "sorry rocks", the name given to the vast number of rocks and sand sent back to the park after being taken away by visitors. This desire to possess the landscape of the park, if only in the form of a rock that is picked up and taken, reflects the strength of this defining settler structure of feeling to claim iconic landscapes such as Uluṟu-Kata Tjuṯa National Park that Mackey (2014, 2016) also discusses in the context of settler-colonial Canada.

A feeling of nationalistic pride in the white possession of Australia, including even its remote desert regions, appeared repeatedly in visitor narratives, such as Naomi's positioning of her visit to the park as a way of obtaining a "deeper sense of understanding of the country that I was born in", with Uluṟu romantically framed as "the heart of our country". The desire to possess Uluṟu was even more striking in the narratives of Vicki and Bill, introduced in Chapter 5, and to some extent also present in my own feelings towards the house I lived in in the Park while conducting my fieldwork, discussed in the text box on the previous page. Vicki and Bill explained that they had eschewed international travel to see more of Australia, because, in Bill's words "I'm Australian, l live in Australia, ... [but] I don't know what's in *my own backyard*" (emphasis added). Talking about other parts of the NT such as Kings Canyon, Bill reflected "... it's so underrated and that's what we tend to do as Australians, we tend to underrate *our own country*" (emphasis added).

Vicki agreed that "... as Australians we don't know what we've got in *our own backyard*" (emphasis added). For Bill and Vicki, the desire to travel in Australia was based, at least in part, on the belief that it is almost a patriotic duty of Australians to know their "own" country by traversing it, evoking the former journeys of white "explorers". In their spoken discourse about their reasons for travel Bill and Vicki described Australia, even locations remote to their normal place of residence such as Uluṟu-Kata Tjuṯa National Park, using possessive pronouns and Australian colloquialisms such as "our own backyard" and "our own country". Collective symbols such as "our own backyard" and phrases such as "our own country" strongly convey a sense of ownership coupled with emotional attachment. Language of this type powerfully evokes Moreton-Robinson's (2015, xii) idea of the "... white possessive logics [that] are operationalized within discourses to circulate sets of meanings about ownership of the nation, as part of common-sense knowledge, decision making, and socially produced conventions". Such "possessive logics" are also very much a settler structure of feeling in which identity and personhood are equated with belonging to the fictive white construct "Australia", which Maddison claims only exists in the "imagination of its political sovereignty" (Maddison 2022, n.p.).

Like visitors such as Jason, who described himself as drawn to "the red dirt country", I was also drawn to this landscape because of its stark contrast to my home in the NSW Blue Mountains. As far as I am aware, my desire to produce art is prompted by a feeling of wonderment or admiration for something observed that is difficult to express in words and is a way of coming into relationship with the materiality of a place, or of authenticating place (Knudsen and Waade 2010). However, part of the appeal of this landscape to me as an artist is that it is alien to my white ways of seeing, and some of the landscape features like the immature desert oaks, shown in Figure 6.1, I found profoundly other-worldly. In the context of the analysis provided in this chapter, I have come to question whether the artworks I produced while living in the park and after my return home express a similar possessive logic and a desire to lay claim to a place that is "other" (Said 1978). By visually representing the park and in this way making it mine, I may be enacting a form of claiming not dissimilar to the narratives of Vicki and Bill summarised above, and perpetuating a continued "othering" of the park as an exotic landscape, labels that may then be inappropriately transferred to Indigenous peoples (tebrakunna country and Lee 2022).

Figure 6.1 Nightwalk, "Rangerville", UKTNP. Photograph by the author of a painting by the author.

Anger, alienation and disengagement at Red Hands Cave

This section explores visitors' affective responses to the interpretive signage associated with Red Hands Cave in the Blue Mountains National Park and, to a more limited extent, the cultural markings within the cave itself. While the interpretation within Uluṟu-Kata Tjuṯa National Park prompted a range of empathic and embodied engagements by visitors, the interpretation associated with Red Hands Cave proved alienating to visitors, and for some, prompted affective responses such as anger. Responses to the interpretation contrasted with reactions to the place itself, primarily the red and yellow hand prints and stencils after which the cave is now named, which inspired feelings of connection, gratitude and wonder. The affective responses of both Indigenous and non-Indigenous visitors to this place points to the role of rock art as "an animate aspect of living landscapes" and a "sentient social agent", that "by its very existence, ...engages the viewer in a relationship with the creators of the art" (Smith 2021, 207). The rock art thus co-creates Country as an affective landscape, one which for non-Indigenous people has poignant echoes of an absent presence, but for Aboriginal peoples evokes connections with ancestors still present at this place.

However, for a number of visitors, the dated interpretation encountered at the cave, and along the walk to it, provided a jarring counterpoint to the arresting cultural heritage of the cave itself. Most visitors linked their negative reactions to the interpretation to the absence of a Traditional Custodian voice or perspective, and in its place a white expert narrator, speaking *about* the culture of Dharug Traditional Custodians in their absence. As discussed in Chapter 3, erasure of a living Indigenous presence within protected areas is typical of older interpretative signage developed within a settler-colonial context, without the input of Indigenous peoples (Ballantyne 1995; Carr 2004; Clarke and Waterton 2015; Porter 2010; Staiff et al. 2002), of which the signage at Red Hands Cave is a further example.

Sue, a Western Sydney local, confided that she visited the cave because she wanted "to share ... [her] passion of culture" with her family, who accompanied her. Her planned visit set her apart from most of the other visitors interviewed at the cave, apart from a group of young Aboriginal women discussed below, who set out to go on a bush walk rather than to experience a place of Aboriginal heritage. This difference in intention points to the way that places of Aboriginal cultural significance in the Blue Mountains were not accorded the same value by settler visitors as a location such as Uluṟu-Kata Tjuṯa National Park, to which visitors deliberately travelled to experience Aṉangu culture. This may relate to a perception of the Blue Mountains National Park as a colonised environment, as it was first traversed by white "explorers" in the late eighteenth century, cumulating in the crossing of the Blue Mountains by Gregory Blaxland, William Wentworth and William Charles Lawson in 1813 (McGrath 2015). It is possible that this perception may have influenced the depth of affective response of settler visitors to places such as Red Hands Cave, which for some, was typified by a shallow or banal register of engagement, which, it has been argued, shuts down a critical dialogic response to heritage places (Smith 2014).

Sue was an exception to this trend, responding with contained anger when I encountered her at the cave and asked her to share her thoughts and feelings on the interpretive signage. She explained with considerable passion,

> I'll tell you what I think when I read the signage everywhere up there (the carpark), and here at the cave, I think how flippin' condescending. I think the way it's, the tense, we're talking about 'them' and 'they'. ...Why is this a colonis[ing] white voice speaking to me about something that has nothing to do with them? Why am I not hearing an Indigenous person's voice? ... It's their message and their story I want to hear when I come here. This, this is an insult. It's an insult to the people that lived here before colonisation. And it's also an insult to my intelligence.

Sue's anger, evident in her tone of voice, breathless delivery and the modality of the phrase "I'll tell you", which left no room for debate, was a powerful affective response to encountering a white rather than Indigenous narrative at a place of Aboriginal cultural significance. Sue's assessment of the interpretation, if not her outrage, was echoed by Tony, a young university student who walked to the cave with a group of friends. Tony reflected that the signage "sounds like it's written by white people … and that it wasn't checked by an Indigenous person or anything like that". While he expressed no anger about this, he commented on how incongruous he found the absence of a Dharug and Gundungurra perspective in the signage, when Western Sydney has one of the largest Aboriginal populations in NSW. Tony also made it clear that he did not find the interpretation informative or engaging and failed to learn anything new about the relationship of Dharug and Gundungurra people with Red Hands Cave by reading it (in fact, the signage focuses exclusively on the Dharug people). Partly because of the hegemonic white discourse position adopted that expressed white ways of seeing as the norm (Erickson 2018), this signage failed to successfully fulfil a pedagogical role and was actually more likely to alienate than engage visitors.

Unlike Tony, whose visit to the cave was incidental, Sara, one of a group of three young Aboriginal women I spoke with at the cave, explained that she and her friends visited because "it makes us feel connected to our culture when we come to places like this. We feel it". Her friend Michelle, who was from the NSW Central Coast, agreed, "I've been to a couple of red hands caves around, like, different Countries. So different people's Countries, so I like to see other people's mob". The third young woman, although she signed a consent form agreeing to participate in the group interview and appeared engaged with what was being said, chose not to contribute verbally. For Sara and Michelle, visiting Red Hands Cave was a way of connecting with their cultural heritage as Aboriginal peoples, and this connection was one they wished to experience relationally, with the other young Aboriginal women accompanying them. Sarah and Michelle's desire to experience this place through direct embodied engagement evokes not only Smith's understanding of heritage sites as "places where people go to feel, to be emotional" (Smith 2014, 125) but also the specific emotional and affective relationships that Aboriginal peoples have with Country that may be felt in the body, and experienced through the senses (Kearney 2009).

In its embodied relationality with place, what Sara described was also an experience of affect, "as a quality of different types of relations through which interrelated bodies are constituted and constantly reshaped … [through] bodily interactions, communications or expectations, as well as positional relations and subject positions in a social and cultural space" (von Scheve 2018, 55). However, rather than being accessible to any visitor, the affective connection articulated by Sara and Michelle was one that was only available to Aboriginal peoples due to their specific relationships to Country (Moreton-Robinson 2022), pointing to the embodied specificity of affect rather than its universality (Tolia-Kelly 2006). By feeling Country, the place itself was also actively "absorbed into an embodied encounter" (Waterton 2019, 97), one in which Country and the body/identity were co-constituting. The landscape of Red Hands Cave is thus best understood as one that was in an active process of co-becoming with Aboriginal visitors such as Sara and Michelle, rather than an inert and static thing. This was also in line with Gundungurra Elder Aunty Leonie's understanding of this place as one that retains its significance because "everything's still there", including the spirits of ancestors (Bird Rose 1996b).

The desire of young Aboriginal peoples living in the Sydney region to strengthen their connection with their cultural heritage by visiting places such as Red Hands Cave can be understood as not only affective but discursive and historic, as it is likely to relate to the wholescale disruption of the relationship of generations of Aboriginal peoples with culture and Country arising from the severe impact of white invasion (Gapps 2018; Karskens 2009, 2020). This disruption occurred in the course of the transformation of Country into settler-colonial space, including through regimes of private property and the creation of protected public spaces such as national parks (Banivanua Mar 2010). The settler-colonial state also removed Aboriginal peoples from Country by relocating them to reserves and missions and took Aboriginal children from their families to be raised by white families and institutions (Armitage 1995).

It was thus doubly distressing when Aboriginal peoples visited places of cultural importance to them only to find that these places, too, had been appropriated by the coloniser (Huggins et al., 1995), producing "encounters with 'self' as 'other'" (Tolia-Kelly 2016, 900) at the very places where they sought cultural connection. At Red Hands Cave, this othering was achieved not only through the text of the signage, which was written for a white audience, but also through the construction of the cave as a museum-like object for viewing. The installation of a metal grate, although intended to protect the cultural markings within the cave, created distance from these. The small windows at eye level located within the grate, designed to look through and take photographs, turned visitors into spectators by emphasising the act of gazing.

While the place itself inspired a feeling of connection, the framing of Red Hands Cave and its interpretative signage detracted from such feelings by misrepresenting the Dharug people in several ways. Sara, for example, objected to the way the signage discussed the Aboriginal occupation of the area in the past tense, referring to it as "now past when it hasn't past at all". Michelle interrupted to agree, asserting that "in reality our culture is still being taught today". Sara passionately elaborated,

> It's still very much alive. And still very strong ... our culture and tribes are still ... very much here. I know a lot of Dharug people that are still living in this area, so they didn't leave here. They still live here. I'm a Wiradjuri woman from Dubbo. And I know a lot of Wiradjuri people still living in that area ...

An emphasis on Aboriginal cultures as living and ongoing was a key discourse strand in the interview with Sara and Michelle and constituted a spontaneously emerging counter-narrative to the myth of Aboriginal extinction constructed in the signage (see Chapter 3). In Sara's view, while Traditional Custodians "cared enough to preserve" places like Red Hands Cave, the NSW NPWS was not adequately fulfilling its responsibility to care for this place because the interpretation was "not being done respectfully, or culturally appropriate[ly]". For her, this reflected generalised disrespectful attitudes to Aboriginal peoples among the Australian community as a whole, "which needs to change ... throughout society". She concluded that the way the signage was written "comes across as we've taken over, white Australia has taken over".

Michelle and Sara's experience at Red Hands Cave calls to mind Tolia-Kelly et al.'s (2017, 3) insight that the "de-historicization [of Indigenous cultural heritage] has situated heritage as reading Aboriginal culture as a prehistoric ... presence in which the 'Other' is not imagined as visiting the site". When those constructed as "other" did visit the site, the text of the interpretation was read as being *about* them in a way that misrepresented Aboriginal peoples and cultures, rather than being *by* them and *for* them in a way that enhanced their connection with culture and place. While Michelle and Sara called for change to this state of affairs, their emotional response to the interpretation and what it represented was one of despondence, neither of them holding out much hope that change was possible. As Tolia-Kelly (2016, 902) explains, "[t]he experience of seeing one's culture through another, results in a veil, a prism of trauma, anxiety and alienation", such that the heritage encounter becomes a "theatre of pain" (Tolia-Kelly 2016, 897) at a place one hoped to experience as a "happy object" (Ahmed 2010, 30).

This painful experience had broader ramifications, as Michelle and Sara extrapolated from the silences of the signage to the broader failure of Australian social and political institutions to acknowledge the impact of white colonisation on Aboriginal peoples. Michelle observed "they don't even teach it [this history] in schools", Sara agreeing "they like to sort of hide that history". Some research participants thus viewed representations of Aboriginal culture at Red Hands Cave as linked to broader issues of white dominance and systematic racism in the Australian context, including a refusal to speak openly and clearly about the reality of European invasion and Aboriginal dispossession. The reality of white racism was again apparent in the racist graffiti on the first sign on the walk to the cave (Figure 3.6),[3] prompting Sara and Michelle to express support for the metal grate that encloses and protects the cave from damage (Figure 3.11). Referring to the grate, Sara responded "I actually like it like that. There's a lot of graffiti over some of the caves we passed. And if this grate wasn't there, I'm telling you now, that would have been destroyed". Michelle agreed "they would have added their names, signatures, anything", Sara concluded, "people have no respect for our culture."

Threat to Aboriginal material culture evoked an affective intensity in Sara and Michelle, apparent in the choice of words such as "destroyed" to describe the nature of this threat. This intensity was also apparent in their tone of voice and conversational pattern, including the high modality of the phrase "I'm telling you now", which left no room for disagreement. There was no pause between Sara's statement and Michelle's conversational follow-up, reflecting an urgency on Michelle's part to speak these words of agreement. As Walkerdine (2013, 759) suggests, the type of affective intensity apparent in Michelle's statement and Sara's response reveals how such intensity is "experienced relationally but not as clearly belonging to one body or other, but more as a flow, flux, or surge within the relational setting or in a group". Emotion, like affect, is not simply "in" the subject or the object but comes into being in relationships between people and things/objects (Ahmed 2004), such as Sarah and Michelle's interaction with each other and the heritage assemblage of Red Hands Cave.

While not banal, Tony's response to Red Hands Cave was definitely in a more neutral register than Sara and Michelle's. Tony, reflecting on his feelings upon arriving at the cave, said he was both "amazed" and "impressed" by the cultural markings within the cave, but at the same time felt "sad to see that there'd been ... a cage put around it". He reported that this made him feel "unhappy ... [that] people can't sort of respect the art, and the culture ... [that] it needs to be shut off to the public to look through the cage. I was disturbed on that front ...". Not only was Tony troubled by the presence of the metal grate around the cave and what it signified, he also explained that he was "cringing" on his way to the Cave, because one of his friends,

> ... is a bit ignorant, let's say, about Indigenous culture and stuff like that. So, he's making all these sort of jokes and the point of me saying that is, I just feel, when people come across these things, they already have ... an idea of Aboriginal and Torres Strait Islander Australia in their head [based, he believed] ... on negative stereotypes.

Tony's description of himself as "cringing" in reaction to his friend's racist comments aptly conveys the affective response of physically drawing into oneself, literally shrinking away from his friend's words, as if to avoid a physical blow. Tony's affective response thus mirrors his reluctance to clearly restate what his friend actually said, instead relying on euphemisms that the friend was "a bit ignorant ... about Indigenous culture". The fact that Tony silently endured the racism of his friend speaks to the affective power of racist discourse in the contemporary Australian context. Nelson (2020) explains that racist performances can be more difficult to challenge in close relationships than in cases of bystander racism because of the subject's investment in maintaining these relationships (Nelson 2020). For this reason, close relationships "are important settings for the performance, reproduction, and transmission of Whites' racist ideas and understandings" (Picca and Feagin 2007, 13).

Visitor responses to Red Hands Cave arose not just from the cultural markings within the cave or their interpretation but involved a level of discomfort prompted by the racism that the material culture of Aboriginal peoples engenders in some Australians. These responses resulted not only from viewing racist graffiti on signage associated with the cave but from the racist talk of companions. The interpretative signage associated with the cave significantly contributed to the discomfort of visitors encountering it, undermining the capacity of some visitors, particularly Aboriginal peoples, to fully engage with this place. The dated nature of this interpretation, the offensive framing of the Dharug connection to this place in the past tense, coupled with the lack of care apparent in the deteriorated state of the signage and its defacement, contributed to the feelings of anger, alienation and despair experienced by some visitors to Red Hands Cave. The heritage discourse and infrastructure at Red Hands Cave thus represent a form of cultural production in which Australian settler-colonial "legal and political mappings serve as the unacknowledged frame" (Rifkin 2013, 330), recreating Aboriginal Country as settler-colonial space (Banivanua Mar 2010).

Sadness at "the weeds taking over": Visitor perspectives from the Jenolan Caves

Like Red Hands Cave, the Jenolan Caves, west of the Blue Mountains, is a landscape that was impacted at a relatively early stage of Australian settler-colonialism. The "discovery" of its spectacular limestone caves led to its transformation into a touristed landscape for the well-to-do of the Victorian age from the late 1890s onwards (NSW Government n.d.b). In addition to its limestone caves, the Jenolan Caves precinct is generally known for the fading grandeur of its colonial architecture, rather than for its Aboriginal cultural heritage. For some visitors who took the Aboriginal Culture and History audio tour on the Jenolan Caves mobile phone app, their encounter with this tour was the first time they were made aware not only of the importance of this landscape to the Gundungurra people but also of the impact of the colonial period on the Aboriginal peoples of this region.

Several visitors were both surprised and angered to learn about the "frontier wars" between local Aboriginal peoples and British settlers and military forces in Bathurst, a regional town approximately 75 kilometres from the Jenolan Caves. The Aboriginal Culture and History tour acquainted visitors with this history, including some of the consequences of the 1824 declaration of martial law in Bathurst. This declaration restricted the mobility of Aboriginal peoples by prohibiting them from congregating in groups of six or more or coming within close proximity to a white settlement (Gapps 2021). Diana, a domestic Australian visitor who lived in Bathurst, was almost speechless with outrage when she learnt of the conditions imposed by the martial law declaration. At the end of the interview, Diana also reflected that she had been shocked and distressed upon first learning of Aboriginal massacre sites in the city of Bathurst. These places, she mused,

> ... places where I've just ignorantly walked through, ... [are] place[s] of a lot of sadness for [Aboriginal] people, and I guess I had no idea how in their face all that kind of history is ... and I feel like there's a bit of an awakening going on and all the rest of it, but actually for the majority of people, there isn't, unless you're reading this stuff and kind of interested in it, Joe Blow down the road is going, 'Well, why don't they just get over it, it happened in the past' and all the rest, but it's actually happening right now, and the trauma [is ongoing].

Confronting a colonised environment, Jenolan Caves, 2019

I spend a weekend at the Jenolan Caves to conduct visitor interviews, surrounded by the fading grandeur of its colonial architecture.

My not so salubrious accommodation is built like a downmarket version of a Swiss chalet. Heritage listed Caves House has fraying carpets and paint peeling from the ornate restaurant ceiling. The artwork in the Café is a collection of sepia photographs of the caves during a very colonial nineteenth century. No reference to the Gundungurra people.

Walking around the Blue Lake, there's evidence of European impact everywhere – the damming of the lake, once a swamp where the Gundungurra fished for eels; pipes and arches; hydroelectric relics, and weeds, a profusion of them.

For Diana, a cultural tour she participated in with a Wiradjuri Traditional Custodian in Bathurst was part of an "awakening", or growing recognition that Australian history is marked by many accounts of grievous harm perpetrated against Aboriginal peoples by white invaders, with ongoing manifestations in the present. Like Eric's response to the hand-back video in the Uluṟu Cultural Centre, discussed earlier in this chapter, Diana also engages in higher-level empathy, through other-orientated perspective-taking (Coplan and Goldie 2011). This history, she learnt, continues to colour the way Aboriginal peoples experience certain places, especially sites of known massacres. However, she was concerned that in her care for this history and how it continues to impact Aboriginal peoples, she was in the minority of Australians. Instead, she believed that most Australians would prefer to regard these events as without ongoing ramifications as they are located in the colonial past. A preference for relegating such injustices to the past, Maddison (2012, 4) reveals, is "one of the most frequently deployed strategies designed to minimise or avoid collective guilt in the present."

Although Diana was already aware that the Traditional Custodians of the Bathurst area had been affected by frontier violence, for Danny, a younger woman from the Blue Mountains, this aspect of Australian history, was "all very new information", something she'd never heard about before. She described her emotional response to learning about this as "sad, angry, frustrated". She also noted the changes to the landscape of the caves such as the construction of the Blue Lake, and wondered about the impact of these changes on the Aboriginal peoples of the Jenolan Caves, exclaiming "That was done in 1908!" [sounds astounded]. Imagine this is where they lived. And then some deuce just rocks up and does that! That would have been horrible". Although the Gundungurra and Wiradjuri peoples did not live permanently at the caves, Danny imaginatively invests in attempting to understand how the transformation of this environment by Europeans may have impacted its Traditional Custodians, also demonstrating higher level empathy (Coplan and Goldie 2011).

The emotional responses of other visitors to their tours on the Jenolan Caves' mobile phone app were also influenced by their perception of the changes that have occurred to the caves since white colonisation. Diana and her mother Ruby reflected throughout their interview that the caves have become a colonised environment, apparent in the European style architecture of many of the buildings, some of which resemble Swiss chalets. During my time at the Jenolan Caves I was also struck by the incongruity of this heritage listed architecture, as I reflect in the text-box on the previous page. Diana and Ruby linked the colonial architecture with the lack of dual naming protocols, whereby places are referred to by both their Traditional Custodian and English names. These observations led Ruby to conclude that,

> ... the footprint [of the caves] seems to have really been settled by white people, to such an extent that it's been eroded. And white people have made a foundation stone here that is, I think, going to be very, very hard to challenge ... You are constantly aware of it ... even to the point that there's some European gardens, and that sort of thing. And ... [it] looks more like a Swiss village.

Ruby felt that you had to "... look past all of that, and you look at where the water's flowing and the rocks and the trees and the bracken and the feel of [it] all" to get a sense of the importance of the caves and reserve to the Gundungurra and Wiradjuri peoples.

Lisa Hill (2015, 420) notes that in "contemporary archaeology, it is recognised that the past is always already present", and that symmetrical archaeology "seeks to uncover pleats and folds in the fabric of time".[4] By looking past the Europeanised built environment, it was almost as if Ruby was seeking to uncover the "pleats and folds" that Hill refers to. Ruby's comments also evoke the affective power of "absent presences" or hauntings, in which the past may be spectral because it has deliberately "disappeared" but can still be felt and imagined. This imagining is not only part of an affective response but a political act because it is a refusal to accept an erasure of the active presence of the Gundungurra and Wiradjuri peoples. Such imaginings, as Hill (2015, 423) explains, citing Derrida (1994, 70), can also function as a form of justice seeking for past generations. Taking into account the continuing presence of the past, Hill (2015, 423) argues that "the concept of the spectral has value because it suggests that our experience of the world is haunted by a space-time in which past and future co-exist, and interact, in uncertain and unpredictable ways". It is this uncertainty and unpredictability that leads absent presences to unsettle, and that perhaps contributed to the feeling of sadness that Ruby and Diana articulated below.

Ruby and Diana expressed a sense of the caves as an environment that had been transformed by Europeans, both intentionally and unintentionally, Diana articulating a "little bit of sadness or disappointment with the weeds taking over". Reflecting on her experience of the walks, the audio-tour and indeed the whole environment of the caves, including passing remnants of historic European heritage in the form of old hydroelectric plant and storm water drains on the walks, Diana reported feeling a "bit sadder", because "… it just shows how much *we've taken away*" (emphasis added). The white "settler" impact on this place, for Diana and Ruby, has detracted from, and diminished, the spirituality of the caves, although this is something they could still sense, as if places have layers of meaning like a palimpsest and Gundungurra Country continued to co-exist, in some way, within or beneath this Europeanised environment. This evokes Yarralin man Daly Pulkara's observation that "this earth has an Aboriginal culture inside" (Daly Pulkara, cited in Bird Rose 1996b, 85) but is also reminiscent of the belief of visitors such as Frank, discussed in Chapter 5, that Uluṟu-Kata Tjuṯa National Park is more spiritual than other parts of Australia because it was colonised later. This feeling of diffuse anxiety about the loss of some original essence or purity within the landscape, thus represented a repeated theme in visitor narratives, evoking not just sorrow, but also white shame and guilt (Slater 2019).

The comment that non-Aboriginal Australians have "taken away" from places like the Jenolan Caves also echoes the deficit discourse apparent in a number of visitors' comments about changes to Aboriginal peoples and cultures since colonisation (Fforde et al. 2013), that could be described as a settler structure of feeling (Rifkin 2011). However, it is not just peoples and cultures that Diana and Ruby include in this assessment, but the landscape also, as if it has been degraded by the impact of colonisation. One way that this is observed is in the impact of introduced plant and animal species on these places. However, based on statements such as Daly Pulkara's, Deborah Bird Rose (1996b, 85) argues that "we must conclude that a holistic world view, which situates life forms as part of living systems, will conceptualise loss in complex and non-absolute ways, and will place confidence in the earth's capacity to recover from damage". While white guilt would tend to construct loss in an absolute sense, the response of Traditional Custodians to changes to Country is often more complex, and may for example, involve a willingness to co-exist with introduced plant and animal species, even invasive ones (Trigger et al. 2008).

Conclusion

In this chapter, I have used the first-hand accounts of visitors to explore the affective, emotional and embodied dimensions of their experiences at recognised places of Aboriginal cultural heritage such as Uluṟu-Kata Tjuṯa National Park, and places that are being reclaimed by Aboriginal peoples, such as Red Hands Cave and the Jenolan Caves. These accounts demonstrate that visitors value forms of interpretation that they can respond to emotionally, and that facilitate embodied, affective and emotive connections with place and/or deeper understandings of the meaning of these places to their Traditional Custodians (Witcomb 2013). It is evident that such experiences can, in turn, change visitors, causing them to value previously undervalued environments and abandon culturally insensitive behaviours, suggesting that visitor responses are influenced by hegemonic colonial discourses but not predetermined by them. A key feature of successful interpretation, such as the audio-tours on the Jenolan Caves mobile phone app, locale-based signage close to Uluṟu in Uluṟu-Kata Tjuṯa National Park and the Uluṟu Cultural Centre, is that it prompted such changes and connections.

A defining feature of successful interpretation in the context of my research is that it has been developed by, or in collaboration with, Traditional Custodians and other Aboriginal peoples, and typically was also delivered by them. This contrasts with dated interpretive signage such as at Red Hands Cave in the Blue Mountains National Park, which failed to engage visitors and, in some cases, served to anger and alienate them. This occurred because Red Hands Cave's signage featured a white "expert" voice and was developed without the involvement of Traditional Custodians. While signage of this type has already been subject to academic critique (Ballantyne 1995; Carr 2004; Staiff et al. 2002), my research contributes to knowledge in this area by exploring visitor responses to it, including those of Aboriginal visitors. For the young Aboriginal women who visited Red Hands Cave whose views are discussed in this chapter, the lack of a Traditional Custodian voice and perspective, and the reference to Traditional Custodians in the past tense, served to confirm their observation that Aboriginal peoples were continually written out of Australian history by the settler-colonial state, with the ongoing epistemic violence of settler-colonialism left unacknowledged (Grosfoguel 2015).

However, more recent interpretive interventions largely controlled by Traditional Custodians, such as the Aboriginal Culture and History tour on the Jenolan Caves' mobile phone app, broach previously prohibited subjects, such as the injustice meted out to the Wiradjuri and Gundungurra peoples during the frontier wars in the Bathurst region. While the emotions prompted by this content were evidently uncomfortable for some visitors, they did not shy away from this challenging material, some welcoming the opportunity to learn a history that they believed had been deliberately concealed. It was sometimes by dealing with uncomfortable emotions such as outrage and distress, concern about their impacts as tourists, and shame about the treatment of Aboriginal peoples under settler-colonialism, that visitors could begin processes of personal transformation with broader social and political ramifications, as both Probyn (2005) and Waitt et al. (2007) have argued.

Shame, when prompting self-examination and abandonment of previous assumptions and beliefs, and when accompanied by an awareness that speaking this shame did not erase it (Ahmed 2004), had the potential to be powerfully transformative for visitors. Furthermore, the capacity for deep empathic engagement with Aboriginal peoples through higher-level empathy (Coplan and Goldie 2011), often through exposure to interpretive self-representation by Aboriginal peoples, could be a significant catalyst for change. However, what Lisa Slater (2019) calls "virtuous anxiety", emerging in visitor narratives of Aboriginal deficits and the need for white intervention to remedy these, not only undermined Aboriginal sovereignty but shut down rather than facilitated visitor transformations. Emotions such as settler pride in the possession of the Australian continent are also best described as "settler structures of feeling" (Rifkin 2011, 350) that reinforce rather than challenge settler-colonialism.

While affective, emotional and embodied engagement can thus be an important component of a progressive politics, certain affects and emotions that arise from and reinforce settler-colonial ways of being are at odds with such a politics (Maddison 2012). By taking embodiment, emotion and affect into account we understand, however, that there is no contained human subject, as such, but rather that being and knowing occur in relationality with other people and the non-human. Although this being and knowing occurs in the moment, it is informed by memory, beliefs and other subjective factors, and is necessarily also a product of settler-colonial discursive formations circulating at social, political and institutional levels (Lähdesmäki 2017; Wetherell 2013). However, recognising that visitor responses were to some extent discursively produced is not in itself a cause for existential pessimism, as discursive formations, as Foucault (2002) has argued, are historically specific and thus subject to change over time. They may also be subject to powerful interruptions by Aboriginal peoples, speaking of the resilience of Country and culture from their own specific nodes of power (Laclau and Mouffe 1985). The transformative capacity of these affective, emotive and embodied encounters with Aboriginal cultures, histories and Countries is a subject taken up in Chapter 7, where I explore the extent to which visitors were changed by these interactions, the form that these changes took and whether such changes persisted over time.

Notes

1 This signage was removed after the closure of the Uluṟu climb in October 2019.
2 In line with principles of research ethics, I checked with Derrick several times about whether he was okay and wished to continue the interview. He was joined by his wife at the end of the interview and the Participant Information Sheet that he was given as part of the consent procedure included the phone number of a confidential counselling service.
3 This graffiti has been present since at least October 2018, when I photographically documented it. Given the deteriorated state of the sign itself, which may not have been replaced since its 1988 installation, it is possible the graffiti has been present for much longer. I reported it to the managing authority in 2019.
4 Symmetrical archaeology challenges the "asymmetric dualisms" (Witmore 2007, 546) characteristic of modernism such as subject/ object, nature/culture, and past/ present. Its key principles include that "[a]rchaeology begins with mixtures, not bifurcation" and "the past is not exclusively past" (Witmore 2007, 549).

7 "Transformations"

Visitor "awakenings" or journeys in "becoming Aboriginal"?

Introduction

In this chapter, I explore the meaning of visitor "transformations" arising from visitors' own accounts of their experiences of Aboriginal cultural heritage in protected areas. Transformative experience can be understood as "the process by which we transform our taken-for-granted frames of reference (meaning perspectives, habits of mind, mind-sets) to make them more inclusive, discriminating, open, emotionally capable of change, and reflective" (Mezirow 2000, 7–8). I am particularly interested in the way that such experiences may be transformative because they challenge visitors' formerly held beliefs and opinions, particularly biased views about Aboriginal peoples, and the extent to which these changes are sustained over time. To a lesser degree, I seek to articulate some of the ways that I as a researcher have also been changed through my engagements with visitors, Traditional Custodians and research locales, in recognition that the researcher is part of the research process.

The chapter begins with an introduction that provides an overview of relevant scholarship in the areas of tourism, visitor subjectivity, and museum and heritage interpretation as an impetus for change or "transformation", before moving on to discuss these issues in the context of visitor interviews. Throughout the chapter, I engage with Aboriginal ontologies and, where relevant, post-human perspectives, to elucidate how some of the changes noted by visitors occurred as a form of "co-becoming" with human and agentic more-than-human worlds or Country. Additionally, I identify exposure to interpretation authored by Traditional Custodians, including Aboriginal perspectives on Australian post-contact history, as a discursive intervention that proved powerfully transformative of some visitors' worldviews. However, such engagements did not occur in a historical, social or political vacuum, but in the context of Australian settler-colonialism, and the broader phenomenon of modernity/colonialism (Quijano 2007), forces which could, and often did, resist and constrain these "transformations".

Although the literature on Indigenous cultural tourism discussed below suggests that visitor engagement with this form of tourism can result in

DOI: 10.4324/9781032654973-7

transformative moments, and deeper and more enduring changes in thinking and behaviour, concerns also flagged in this literature were borne out in my own research. These concerns relate to the limited extent that visitors can reasonably be expected to transform on the basis of one-off, relatively short cultural tourism experiences. This is particularly the case, when, as discussed in Chapter 5, discourses that construct Aboriginal identity as deficit based, and plagued by intractable problems, pervade Australian settler-colonialism (Fforde et al. 2013; Fredericks 2013; Gorringe et al. 2011). Although accounts of visitors themselves reveal some very real shifts in perception and behaviour, for example, learning to see iconic national landscapes like Uluṟu-Kata Tjuṯa National Park as Country, with unique meaning to A<u>n</u>angu, other visitors, such as those interviewed at the former climb site, were fixed in their views and refused such insights. The chapter concludes with a discussion of the limitations in the "transformations" experienced by those visitors who were seeking an engagement with Aboriginal peoples, cultures and landscapes. While some made important progress in changing their "habits of mind" (Mezirow 2000, 7–8), many also retained certain settler-colonial assumptions, suggesting that such visitor "transformations", perhaps inevitably, are partial rather than complete.

Understanding tourism, postcoloniality and "transformation"

Key questions explored in this research are whether visitors to places of recognised Aboriginal cultural heritage can develop a (self-identified) connection with Country and a deeper self-perceived understanding of Aboriginal cultural heritage. Relatedly, I have also asked what type of interpretive experiences may facilitate this. I thus hypothesised that visitors might experience processes of change, or what Bawaka Country et al. (2015, 270) call "an ontology of co-becoming", when they visited protected areas, recognising the agentic nature of Country itself in these hypothetical transformations.

The idea of travel as a potentially transformative experience is almost a cliché in writings on tourism and is a trope in tourism advertising, with the potential to engender desire that the actual tourist experience may leave unsatisfied (Bruner 1991; Lean et al. 2014; Smith 1978; Stronza 2001). As Garth Lean (2016, 2) observes,

> [t]ravel remains heavily promoted as an agent of change … it is often claimed to promote learning (e.g. of languages, cultures, history, religions and places) … cosmopolitanism, personal enlightenment, cross-cultural understanding, an awareness of various global issues, environmental consciousness and 'wellness'.

However, as far as Indigenous cultural tourism is concerned, there is a danger that the cross-cultural understanding gained will be minimal and that it will fail to challenge embedded assumptions and stereotypes. One example of the way that tourism can perpetuate settler-colonial discourses is through travel to Australia's "outback" regions as a nationalistic "rite of passage" (Crang 2005, 37), including the pilgrimage of visitors to Uluṟu-Kata Tjuṯa National Park to climb Uluṟu, prior to the climb's permanent closure in October 2019 (Smith 2012; Whittington and Waterton 2021).

The transformative nature of tourism is thus contested on the basis that it can reinforce what Lean (2016, 2) describes as "existing way of seeing and acting in the world, supporting prejudices, misguided/'false' representations, romantic visions/performances and ... the continuation of colonial relations ...". These insights are particularly relevant to the participation of domestic and international visitors in Aboriginal cultural tourism in settler-colonial nations such as Australia. In this context, the type of "transformations" that visitors experience may actually be a form of cultural appropriation of Aboriginal spirituality and even identity (Galliford 2012). However, Mark Galliford (2012) distinguishes shallow forms of engagement based on cultural stereotypes from the "transformations" that can arise from,

> having been affected through intimate experiences with Aboriginal people, an opening up to difference- becoming-Aboriginal- that has impacted on my life and sensibilities, and opened out ... a more respectful and aware sense of my own singular, yet national, presence.

Although the changed thinking Galliford (2012) describes in visitors and himself was quite profound, extending to understanding aspects of Aboriginal ontologies of Country, he also argues in favour of visitor Indigenisation as a way of cementing the belonging of non-Indigenous people in post-colonial Australia. I take issue with this argument, including its use of terms such as "becoming-Aboriginal" as a metaphor for visitors' co-becomings with Aboriginal peoples and places, as an example of the type of appropriation Galliford (2012) also cautions against.

The connection between tourism and coloniality is not a new theme in the tourism literature and in postcolonising Australia can also be read in attempts to colonise Aboriginal identities. Australian and international scholarship on travel to places recognised as Indigenous or territories of former colonial powers by minority world travellers, including recent work focusing on domestic tourism within settler-colonies, has highlighted the many linkages between tourism and colonialism (Grimwood et al. 2019; Hall and Tucker 2004; Higgins-Desbiolles et al. 2022; Linehan et al. 2020; Palmer 1994; Pratt 1992; Urry 2011; Walter 2021). Earlier work such as that by John Urry (2011), critiqued travel by minority world elites for seeking out the exotic non-Western "other", the binary of the modern white Western male subject, who was inevitably viewed in an objectifying way through the "tourist gaze". Such "othering" remains apparent in the discourses of visitors participating in my research, such as Eve and Katrine, who took part in the Blue Mountains Walkabout Tour (see Chapter 5 and later in this chapter).

Travel to former minority world colonies such as the Caribbean by citizens of those very colonial powers has also been criticised for romanticising and reinforcing neo-colonial power relations (Palmer 1994). Similarly, more recent work in this field has focused on travel within settler colonial nations such as Canada and the United States as enacting settler-colonial power dynamics and reifying stereotypes of Indigenous peoples (Grimwood et al. 2019; Walter 2021). While many of these observations are pertinent to my research, I also found that visitors to protected areas, given their embodied diversity in terms of age, gender, and nationality, for example, could not easily be summed up with reference to a single tourist gaze. Some visitors immediately challenged common stereotypes, while the way that others perceived Aboriginal peoples and places was subject to marked change as a result of their experiences.

It is also the case that Aboriginal-run tourism enterprises can disrupt stereotypes and other misconceptions. This can occur by Aboriginal peoples taking control of their own representations through the messages that are delivered to tourists (Galliford 2010, 2012; Travesi 2018). Based on her ethnographic work with visitors to Bardi and Jawi Country in Western Australia, Celine Travesi (2018, 278) reveals that the role of Traditional Custodians as tour guides allowed visitors to recognise "Aboriginal people as holders of knowledge (as knowledgeable)" and "to represent themselves as authoritative". By enacting their "ecological knowledge", they were able to "represent themselves, and possibly through this, Bardi and Jawi people more generally, as knowledgeable and capable people" (Travesi 2018, 282), and so challenge deficit discourses about Aboriginal peoples. Aboriginal cultural tourism is thus one way for Aboriginal peoples to disrupt colonial power relations and at the same time assert their right to represent themselves "authentically" (Travesi 2018, 283), on their own terms.

Related scholarship suggests that it is not just tour content but the opportunity afforded by Aboriginal cultural tours "to meet Aboriginal people and share some degree of intimacy with them" (Galliford 2010, 230) that has potential to change the way that visitors think about Aboriginal peoples (see also Walker and Moscardo 2016). Because of Aboriginal cultural tourism's capacity to break down stereotypes and build greater respect for Aboriginal peoples, it is recognised as having a role to play in galvanising social change movements such as reconciliation between Aboriginal and non-Aboriginal Australians (Galliford 2010, 2012). Lloyd et al. note that such tourism,

> may engender transformations that support social justice precisely because they are premised around 'encounters across difference'... [and] their emotional, affective, and sensory aspects—have the potential to unsettle established habits of thought and to open new ethical and moral relations between peoples and places"
>
> (Lloyd et al. 2015, 2)

Engaging Jack Mezirow's (2000) transformative learning theory, Lloyd et al. (2015) argue that it is the unsettling and disorientating nature of such experiences, involving dilemmas which disrupt taken for granted worldviews and ways of thinking, followed by reflection and discussion, that can lead to "an enduring shift in one's habits of thought and perspective" (Lloyd et al. 2015, 4). Similarly, Galliford argues that simply being "on Aboriginal land ... surrounded by Aboriginal signs of a vast historical past and a strong ongoing presence" can problematise or decentre "taken-for-granted notions of national identity and belonging" (Galliford 2010, 232).

Galliford (2010, 234) notes that Freya Higgins-Desbiolles (2003) "has formally defined this particular style of intercultural tourism as Reconciliation Tourism", a form of tourism involving "experiences that educate and foster understanding, contribute to healing relationships ... encourage the development of 'bridges', and help deliver justice to Indigenous people" (Galliford 2010, 234). In order to facilitate this, he argues that social and political elements should be incorporated into Aboriginal cultural tours, rather than simply reinscribing versions of an authentic traditional Aboriginality which can reinforce "preconceived perceptions and stereotypes" (Galliford 2010, 235). The way in which the A<u>n</u>angu host of the Patji Cultural Tour, discussed in Chapter 4, included storytelling about his people's land rights struggle, and his own family's experience of the Stolen Generations, is an example of content of this type.

Kaye Walker and Gianna Moscardo (2016, 1243) suggest that transformative visitor outcomes are those that "increase awareness and create long-lasting impacts through powerful affective, cognitive and behavioural responses". Affective, embodied experiences such as visitor involvement in cultural tours, may be more likely to achieve these outcomes than exposure to on-site signage because they do not simply present information. As Andrea Witcomb (2013, 269) argues, referring to the work of Benjamin (1979), "... information renders the critical faculty inactive [and] ... does not encourage deep attention", in the same way as experiences that engage the imagination and the senses. However, Bawaka Country et al. (2017), writing about the experiences of visitors to Bawaka Cultural Enterprises in North East Arnhem Land, while emphasising the dramatic changes in the ways that some visitors thought about Aboriginal peoples and responded to racist encounters, also identified a number of limitations. They make the point that there are limits on the extent to which non-Aboriginal people can be transformed by short encounters with Aboriginal peoples in place, arguing that "in the complex world of (post)colonial Australia, any notion that a non-Aboriginal person could be simply and statically 'transformed' would be, at best, simplistic and, more likely, a recipe for a non-reflective racism" (Bawaka Country et al. 2017, 461). They also warn of the

> danger of overplaying the transformation of world views that these visitors experienced [because] ... certain messages may be ambiguous and

> false dichotomies may remain unchallenged or untransformed ... supporting romantic stereotypes in which the Bawaka experience was seen as more authentic than the experiences of Aboriginal people living in more colonised milieus.
>
> (Bawaka Country et al. 2017, 461)

Thoroughgoing transformation, as both Bawaka Country et al. (2017) and Smith and Campbell (2015) note, must also extend beyond one tourism encounter, and manifest in continuing attitudinal and behavioural change rather than dissipate over time. Ongoing change may, for example, be apparent in visitors seeking to maintain relationships they have established with Aboriginal communities and in the formation of new ones, as discussed by Kate Lloyd et al. (2015). Some of my participants expressed a similar desire to continue engaging with the Aboriginal peoples who offered the Yellomundee Culture Camp, by returning to assist the Landcare group looking after that place. Despite encouraging findings about visitor intentions to continue engaging with Aboriginal peoples, other research findings discussed below suggest that the cautions of Bawaka Country et al. (2017) are well warranted. I have chosen to premise the discussion of visitor self-identified learning and change in this way to foreshadow my own reservations with the concept of tourist "transformations" and my understanding that challenging racism is an ongoing process, for which visitors are themselves responsible.

Visitor self-identified "transformations"

The discussion in this section is based in part on visitor responses to interview questions that asked them what, if anything, they had learnt during their time in protected areas, and if anything had changed for them as a result of their experiences. Visitor responses to other interview questions relevant to the topic of visitor "transformations" are also drawn upon, as are data from follow-up interviews with four cultural tour participants undertaken three to four months after their travel.[1] The purpose of the follow-up interviews was to assess if any change in attitude or behaviour had been sustained or progressed from the time the initial interview took place. The themes explored below include the perspectival and value shifts some visitors experienced as a result of exposure to Aboriginal interpretations of post-contact Australian history, the emergence or strengthening of feelings of respect and admiration for Aboriginal peoples, changed understandings of place and ways of seeing the landscape, and a sense that they, as non-Aboriginal visitors, had also developed a relationship with Country. For some participants, particularly those from the Yellomundee Culture Camp, increased awareness of Aboriginal ontologies of relationality, time and Country was a further outcome (Brigg and Graham 2020).

Responses to Aboriginal perspectives on post-contact history

Visitors to both Uluṟu-Kata Tjuṯa National Park and the Jenolan Caves responded to learning of the post-contact history of the Traditional Custodians of these places in affective, emotional and cognitive ways, with the engagement of all of these capacities playing a role in transformational change. The "hand-back" video playing in the Uluṟu Cultural Centre at the time I was conducing my research was an engaging firsthand account from Aṉangu about how they felt about "hand-back" and why the return of their land was important to them. Rather than a didactic, information heavy presentation, the style of the video foregrounded the voices and personal narratives of Aṉangu, and for this reason was very effective in engaging visitors (Witcomb 2013). For visitors like Derrick and Eric (introduced in Chapter 6), the video helped build empathy for Aṉangu, however, Jennifer's response (also introduced in Chapter 6) was more self-centric. As discussed in Chapter 6, the pride she articulated about "hand-back" was likely to stymy any change in perspective, shutting down the potential for behaviour change by resolving feelings of discomfort before meaningful change had been achieved (Ahmed 2005; Probyn 2005; Smith and Campbell 2015). It is thus important to recognise that feeling-based responses, while they have the potential to engender "transformation", are not necessarily always positive or progressive (Kidd and Sayner 2019).

In contrast, Eric's reaction to the "hand-back" video was characterised by "higher-level empathy" (Coplan and Goldie 2011, xxxiv), involving imaginative investment and perspective-shifting, leaving him with feelings of increased respect for Aṉangu and their land rights struggle. Eric's reaction reveals that the inclusion of social and political content such as the history of park "hand-back" in interpretive material, as argued by Galliford (2010), can effect positive changes in the way that visitors think and feel towards Aboriginal peoples. This is especially so when affective and emotional reactions also incorporate perspective-shifting (Coplan and Goldie 2011), assisted, in Eric's case, by being able to articulate how and why his perspective had changed. Comparing responses of domestic Australian visitors such as Jennifer with that of Irish national Eric also suggests that international visitors may be less invested in the settler-colonial politics of Australia, and thus more able to experience the "hand-back" video in less settler-focused ways.

Unlike Uluṟu-Kata Tjuṯa National Park, associated as it is with nationalistic journeys of settler-colonial pilgrimage, visitors to research locales such as the Jenolan Caves were less invested in redeeming a settler-colonial identity through emotions such as pride. Although they also valued learning about the post-contact history of Western Sydney, the Blue Mountains and the Bathurst region, they found accounts of the "frontier wars" confronting and distressing rather than affirming or uplifting. However, these uncomfortable feelings could also be productive (Lloyd et al. 2015; Mezirow 2000), in that they led some visitors to re-interpret familiar landscapes based on their post-contact histories, seeing these places afresh. For example, Danny, introduced in Chapter 6, who took the Aboriginal Culture and History Tour on the Jenolan Caves mobile phone app, learnt about the frontier wars for the first time through the tour. As she traversed the landscape of the Jenolan Caves Reserve while listening to the audio tour, Danny worked to make sense of what she had learnt in the context of this landscape, noticing the hydroelectric relics and the Blue Lake itself as evidence of settler-colonial encroachment on a place of great significance to the Gundungurra people. Learning this history while actively moving through a place that was proximal to the frontier wars meant that she perceived the landscape through the lens of this history, changing the way that she understood this place.

While Diana and her mother, Ruby, introduced in Chapter 6, also took the Aboriginal Culture and History Tour, unlike Danny, they were already aware of the frontier wars and Aboriginal massacres that occurred in Australia from the beginning of white settlement up to the 1930s (Gapps 2018, 2021; Ryan et al. n.d.). However, prior to taking the mobile phone tour they did not know these wars impacted the Gundungurra, nor were they aware of the 1824 declaration of martial law in Bathurst. Diana worked to produce meaning from her engagement with the mobile phone tour, integrating this new information with her existing knowledge and experience (Fraser 2007) that was based in part on having previously taken a cultural tour with a Wiradjuri Traditional Custodian in Bathurst. Learning about massacres of the Wiradjuri prompted a feeling of deep empathy for the Wiradjuri people, which Diana was then able to apply to the Gundungurra, an example of meaningful pedagogy leading to learning transfer (Van Winkle and Backman 2011).

Diana reflected on the difficulty faced by Aboriginal peoples living with the many reminders of the historical figures implicated in colonial violence in the town of Bathurst, such as Governor Macquarie, after whom a street is named, and the surveyor George Evans, credited with the discovery of the Bathurst Plains and memorialised via a prominent statue in the Bathurst town centre (Whittington 2024). For Diana and her mother Ruby, learning about the colonial history of Bathurst and the Jenolan Caves catalysed an "awakening" to the ongoing manifestations of colonialism in their hometown and broader regional environment, identifying it as the type of transformational experience discussed by Mezirow (2000). Diana also realised that the harm of colonialism is perpetuated in the present by other Australians denying the importance of this history and its ongoing ramifications. The changed thinking that Diana describes was prompted by exposure to the Gundungurra perspective on Australian post-contact history that the tour provides, but extrapolated from this based on prior knowledge and experience to conclude that the violence of settler-colonialism in Australia is ongoing. Transformational learning may not just be the result of a single experience but can be cumulative, as Fraser (2007) notes. Meaning-making of this type, where conclusions are drawn about broader Australian social justice issues related to Aboriginal peoples, is an important outcome of Aboriginal cultural tourism because of its potential to further meaningful reconciliation between Aboriginal and non-Aboriginal Australians (Galliford 2010).

Altered understandings of Aboriginal cultures and Countries

While learning about post-contact Australian history from the perspective of Aboriginal peoples was to some extent uncomfortable knowledge for some settler Australians, encounters that deepened their understanding of Aboriginal beliefs and lifeways did not have the same associations. In Uluṟu-Kata Tjuṯa National Park, the most common form of learning and change that visitors identified was obtaining knowledge of Aṉangu lifeways (past and continuing), law/lore, and creation accounts. These are all components of Tjukurpa, however, no visitor mentioned this term by name, despite being introduced to it via the various forms of interpretation offered in the park, suggesting they stopped short of grasping this Aṉangu ontology. The new knowledge they gained was associated with one of the most common changes identified by visitors, that of gaining a broader understanding of Aboriginal cultures, including those of Aṉangu and the Dharug and Gundungurra peoples of NSW. This growing understanding was sometimes accompanied by a deeper respect for Aboriginal peoples and their cultures, including for their ingenuity in developing effective ways of living in harsh environments such as Central Australia and for the strength and perseverance demonstrated through their land rights struggles. Increased visitors' respect for Aboriginal peoples as a result of participation in cultural tourism were also outcomes identified in empirical research by Galliford (2010) and Lloyd et al. (2015).

For example, Irish national, Eric, mentioned earlier in the chapter, reflected that the arid landscape of Uluṟu-Kata Tjuṯa National Park prompted a feeling of "respect ... that there's a civilisation that doesn't need Western amenities to live and thrive in that environment", concluding that "they've obviously got something right, in terms of how you only take what you need ... [and] you can live sustainably off the land by being responsible". If learning is understood as a form of meaning-making actively produced by visitors as a response to interpretation encountered in place (Watson et al. 2012), the ensuing transformational change for some visitors was to view traditional Aboriginal lifeways in increasingly positive terms. However, this positive assessment was not equally applied to all Aboriginal peoples, with visitors to Uluṟu-Kata Tjuṯa National Park such as Jason and Vicki attributing deficits to urbanised Aboriginal peoples (see Chapter 5). Jason and Vicki's responses validate the concerns expressed above by Bawaka Country et al. (2017), and also articulated by Lean (2016), that tourist engagements can reproduce colonial relations, particularly in a postcolonial context. Smith's (2014) argument that travel to heritage places simply reinforces what visitors already know, feel and believe seemed to aptly apply to these visitors.

However, other visitors were able to articulate marked changes in their attitudes and perceptions towards both Aboriginal peoples and Country.

Seeing the grid, February 2022

On the way up on my final trip to Uluru, three years on, I look out of the plane window and see the way the rural landscape below is organised into the ordered squares of modern agriculture.

I immediately recognise this as a way that Australia has been transformed into settler-colonial space (after the late, great, Tracy Banivanua Mar) without even having to think about it.

The landscape changes again the closer I get to Uluru – there are less obvious signs of repetitive white geometries and more seeming randomness, in the haphazard rambling of vast inland waterways that continue to mark the landscape, in flood or drought.

The increased knowledge and understanding of Aboriginal peoples and their cultures identified by some visitors was often also linked to a changed understanding of protected area landscapes. For example, a number of visitors to Uluru-Kata Tjuta National Park reported coming to understand how marks and surface formations on the Uluru monolith were associated with Dreaming accounts. Ranger guided walk participant, Bill, introduced in Chapter 5, emphasised that this experience had led him to see Uluru as more than "just a big rock in the middle of nowhere", noting, "... it was fabulous to understand what the rock was about and the different caves". Bill elaborated that participating in the walk led him to understand "why there are certain parts of Uluru marked the way it is".

While this might seem like a minimal level of understanding, it is significant because it demonstrates a shift in perception from viewing Uluru simply as a landscape important for its aesthetic value or as an Australian icon, to beginning to see it as Anangu do, as Country, replete with stories, meaning and powerful non-human others (Bowman 2015). As I observe in the above text-box, my ways of seeing also changed as a result of this research, coming to recognise that the Australian landscape, even remote locations, has been transformed by settler-colonialism. For Bill, learning to see the landscape, however, altered, as Aboriginal Country, could be viewed as decolonising as it challenges a neo-colonial view of landscape as a consumable resource available for settler-colonials, without sentience or agency (Howitt et al. 2013).

By moving through the landscape and observing the signage while listening to the voice of the guide, Bill was also engaging in a process of multisense-based meaning making and co-becoming with other visitors and the non-human (Bawaka Country et al. 2015). The ensuing transformation in his way of seeing made it appear as if the landscape itself had also changed (see also Waterton 2019). However, giving credence to the primacy of Tjukurpa limits the extent to which Country could be considered to co-become with visitors like

Bill, as this A_n_angu ontology recognises the meanings of place that inhere in creation accounts as fixed in the timeless now (James 2015). Other aspects of being in place like the season, the weather, the presence or absence of visitors, the behaviour of plants, animals and insects and even events such as the closure of the Ulu_r_u climb, point to the dynamic nature of place, but A_n_angu ontologies insist that Tjukurpa is enduring. The shift in perception that Bill describes was to recognise the primacy of Tjukurpa in this landscape, so that the land was acknowledged as an A_n_angu place. Particular aspects of the landscape, such as the Ulu_r_u monolith, were thus understood for their significance to A_n_angu, rather than for any collective cultural meaning that Ulu_r_u might also have to settler Australians like himself. The ranger guided walk thus successfully communicated to Bill the hidden meanings attached to Ulu_r_u, foregrounding an ontology that is labelled unimportant and peripheral under settler colonialism (Dussel 2012)

Vicki, another ranger guided walk participant, discussed how she would take the new knowledge obtained on the walk forward in her life, suggesting that "we (non-Aboriginal Australians) are becoming better custodians by Traditional Custodians sharing their knowledge with us". Vicki's willingness to identify herself as a custodian shows that she had been sufficiently influenced by the Aboriginal ontology of caring for Country to integrate it into her self-concept, suggesting that this is one outcome of coming to better understand an Aboriginal worldview. Vicki's adoption of the positionality of custodian could be regarded as a formerly "devalued" Aboriginal ontology (Grosfoguel 2015, 42), that of custodianship or care for Country, gaining broader traction, and beginning to influence the thinking of non-Aboriginal Australians.

However, I argue that by claiming the identity of custodian for herself Vicki was in danger of cultural appropriation. Her repeated assertions that visitors were able to become custodians during our interview, while possibly intended as a form of respectful imitation, was also a settler-colonial appropriation of an Aboriginal way of being that she did not fully understand. Her adoption of this identity was not part of the organic process that Harrison (2014, 47) describes as occurring between Aboriginal and non-Aboriginal workers in the NSW pastoral industry, which leads him to identify a "relationship between pastoral narrative forms and Aboriginal songlines", for example. Harrison (2014, 51) concluded that "the archaeology of shared histories is thus an archaeology of *becoming* in which categories of colonized and colonizer must necessarily be troubled". While this may occur as a result of the entanglements arising from close, lived relationships between Aboriginal and non-Aboriginal people, the limited engagement with Aboriginal peoples arising from Vicki's NT tourism encounters did not disrupt these categories. Rather, her application of this term to herself could be seen as a manifestation of what Wolfe (2006, 387) has termed a "logic of elimination". This involves non-Aboriginal Australians adopting Aboriginal lifeways with a view to replacing Aboriginal peoples, or at the very least, legitimising their own claim to Australia (Horáková 2022). This reading is in line with Vicki's nationalistic pride in the landscape of Uluṟu-Kata Tjuṯa National Park discussed in Chapter 6.

In claiming the identity of custodian, Vicki was also oblivious to the fact that Aboriginal ontologies such as a relationship with Country are integral to Aboriginal sovereignty (Moreton-Robinson 2022; Mutu 2022). While settlers like Vicki may understandably wish to live in a less destructive relationship with their environment, this does not mean that Aboriginal lifeways are fully open to them or that they can be unreflectively deployed as strategies for securing settler futures, for example, in the face of environmental catastrophe (Whyte 2022). Aboriginal sovereignty thus places limits on the extent to which visitors can be "transformed" by their experiences, when these transformations cross a line into appropriating Aboriginal identities and ways of being.

Although some visitors "transformations" led them to claim certain romanticised versions of Aboriginality as part of their own identity, the changes experienced by other visitors are better described as decolonising. For example, a broad range of interpretation that visitors engaged within Uluṟu-Kata Tjuṯa National Park led to new insights about the cultural meanings of the landscape. As Anne explained when I encountered her on the Mala walk, the interpretive signage "makes a layer of significance legible that isn't otherwise there. It makes it a more meaningful experience". Similarly, Ben, after reading the signage at the base of Uluṟu, believed that he'd gained an "appreciation of Country and the significance of the place to the Indigenous people". While such learning was inevitably superficial, due to the limited knowledge able to be obtained from interpretative signage alone, when accompanied by a guided cultural tour, a deeper level of engagement was possible. For example, significant visitor shifts in perception and understanding occurred on the Yellomundee Culture Camp and on the Patji tour, including when our A<u>n</u>angu host communicated restrictions on sharing certain cultural knowledge with visitors.

Arthur, introduced in Chapter 6, learnt that what to him were simply interesting landscape features, the rocky outcrops that from time to time broke up the flat, seemingly featureless landscape to the south of the national park, had stories, but our A<u>n</u>angu guide informed him that these "could not be told". Learning that A<u>n</u>angu knowledge associated with certain places was not only unable to be shared with non-A<u>n</u>angu but was also gender specific, added a further layer to his awareness of the landscape, one that was outside of Eurocentric epistemologies. The restrictions on knowledge that exist within Aboriginal cultures place a further limitation on the capacity of non-Aboriginal people to access and share in such ontologies, and consequently to "care as Country", as advocated by Bawaka Country et al. (2013, 185). By informing Arthur of this knowledge that was held exclusively by A<u>n</u>angu, our guide was also demonstrating that he was an authoritative and knowledgeable person, with responsibilities under Tjukurpa for protecting this knowledge (Travesi 2018).

The new ways of seeing articulated by Arthur and Bill did not occur only at places such as Uluṟu-Kata Tjuṯa National Park or surrounding APY lands, which are broadly recognised as Aboriginal places but also in protected areas close to the Sydney urban interface such as Yellomundee Regional Park and the Jenolan Caves. These shifts in understanding were perhaps more significant among NSW research participants due to the lack of recognition of these peri-urban research locales as having ongoing meaning to their Traditional Custodians (Porter 2018). This was especially true for the Jenolan Caves, which is imbued with a strong European atmosphere by its chalet-style architecture and traditional European gardens, with no obvious material traces of Aboriginal past presence accessible to visitors. However, taking the Healing Waters tour on the Jenolan Caves mobile phone app prompted Suzi, a Chinese international student, to observe, "I'm more knowledgeable and inform[ed] about the place, the place has more meaning than just a cave or water for me". Similarly, Blue Mountains local Danny found that the tour "made me look at it [the Jenolan Caves] in a different way, not knowing anything about it being such a special place. It really has kind of opened my eyes to all that".

Diana, who also took the Healing Waters tour, was interested to learn that the waters flowing through the Caves had been a healing place for the Gundungurra, observing "you could get a really good visual of people using or living in that place or bringing people to that place". Similarly, Carol, a domestic visitor from outer Western Sydney, reflected that hearing the Gurrangatch and Mirrigan creation account "sort of made me have a greater connection to the surroundings, to learn … that's how the river carved it's way". In this creation account, it was a battle between these two creator beings that shaped the Jenolan River running through the Jenolan Caves Reserve. Coming to appreciate the significance of the caves to the Gundungurra and the history of its use by them prior to white invasion, imbued the place with deeper meaning for these visitors, particularly landscape features such as the Jenolan River that are prominent in Gundungurra creation accounts. The caves thus resembled a palimpsest with layers of meaning that revealed themselves to visitors through the audio tour, facilitating a deeper level of engagement with the landscape and a new understanding of the Jenolan Caves as an Aboriginal place. The Gundungurra perspective encountered on the Healing Waters tour thus enabled some visitors to transform their "taken-for-granted frames of reference" (Mezirow 2000, 7–8) to incorporate Gundungurra understanding of this (cultural) landscape.

However, this new appreciation also brought into sharp focus the profound changes to the Jenolan Caves arising from colonisation. As discussed in Chapter 6, Diana and Ruby reflected that the Jenolan Caves had become a Europeanised environment, arguing that the transformation of this landscape by settler Australians would be, for this reason, very difficult to challenge.

My conversation with Diana and Ruby had a strong impact on me. As I drove home from the Caves, passing the upper Blue Mountains townships of Blackheath, Medlow Bath and Katoomba, I reflected that these townships were also Europeanised by nineteenth-century tourism in a similar way to the Jenolan Caves. Echoing Gelder and Jacobs' (1998) idea of the postcolonial uncanny, I began to see a place I understood as home as a colonised environment, no longer taken-for-granted, safe and familiar but alien and unsettling, with all of the uncomfortable emotions associated with this. As Hill (2015, 422) notes, "[t]he power of the uncanny is arrived at not by an encounter with anything strikingly odd or unknown; rather, it is held within something that has a familiarity about it".

Familiar architectural landmarks like The Hydro Majestic Hotel, in Medlow Bath, one of the oldest hotels in the Blue Mountains, which I had previously taken for granted as part of the landscape, I now saw as a marker of settler-colonialism. Conducting research in a place I regarded as home thus also led to a shift in perception of my own environment, accompanied by the affective responses of discomfiture and unsettlement (Plater 2017). Visitor responses thus prompted changes in my own thinking, demonstrating the way that critical qualitative inquiry can be a reciprocal process that also changes the researcher, and that these transformations inevitably have an affective dimension. This reciprocity reflects the way that I was not fully separate from my interviewees or the places that we experienced together, since subjectivity can be understood as "transveral [and]… not restricted to bound individuals" (Braidotti 2019, 33), and emotions and affects also exceed bodily boundaries (Ahmed 2004; Bondi et al. 2007).

Understanding and applying Aboriginal ontologies

Like the mobile phone tours taken by visitors to the Jenolan Caves, the week-long Culture Camp at Yellomundee Regional Park also resulted in participants coming to understand this peri-urban landscape as an Aboriginal place, indeed as part of Country. The implication of recognising Yellomundee as Country is that the whole of Australia, including landscapes that have been drastically altered by colonisation, must also been viewed as unceded Aboriginal Country, as recognised by Porter (2018) and Fredericks (2013). Yellomundee research participants' modelling their relationships with Yellomundee Regional Park and each other on those of the Traditional Custodians who led the Culture Camp may thus have been a way for this group of non-Aboriginal Australians to attempt to come into right relations with Aboriginal peoples and Country. Each of the four Yellomundee Culture Camp participants I interviewed believed they had developed a deeper connection to the landscape of Yellomundee during their time at the camp, coming to view it as an agentic being with whom they could have a relationship, and in this way decentring human authority as discussed by Bawaka Country et al. (2015). They described how they had learnt to listen to Country, to give back rather than only take from Country, to be more aware of other beings, human and non-human, that were also part of Country, and to try to value or re-use parts of Country that might otherwise be discarded, such as plants defined as weeds. This learning led to a significant change in participants' ways of thinking: the ontologies through which they understood the world.

Andrew explained this changed thinking as arising from,

> ... a very profound learning experience. Just feel really lucky to have Uncle Lex and Uncle (name withheld). I feel like it, every sort of moment that I've had with them, I've always had some sort of learning experience, or they've said something that's sort of tweaked with me and been like, well, 'that's amazing', or ... 'that's a way of thinking that I hadn't really been exposed to before'.

Being able to spend time, converse and interact with the Aboriginal peoples who led the camp over the period of one week was thus, as Galliford (2010) has argued, itself powerfully transformative. Andrew's description of the Culture Camp models Mezirow's (2000, 7) principles of transformative learning set out in the introduction to this chapter, in that his approach to his experience is open and reflective and willing to question "taken-for-granted frames of reference". While Andrew emphasised changes to his way of thinking, for Simon the Culture Camp created a strong bond with the place. Simon elaborated,

> I think just with this whole Culture Camp, it's kind of helped me probably have a deeper connection to a place here, in such a short period of time than I probably had anywhere else, maybe apart from my home ... but just learning about the way that the trees interact, the way the plants interact with each other, the animals interact with these. And all that kind of thing just seems to deepen my connection to this place, rather than just a superficial connection of camping here.

When asked specifically if he felt like he had developed a self-perceived connection to Country during his stay at Yellomundee, Simon answered in the affirmative, explaining,

> I feel like it's something hard to describe. I think definitely a part of it has kind of been the idea of taking, but also giving back. So, I mean, like, I feel like we [have] taken from Country here, whether it be like the food which might not necessarily be from here. It's still from Country. But we're also giving back by ... all the weeding up there ... [it becomes a] reciprocal relationship ... it definitely feels like it's created a connection and a relationship with Country.

Andrew described this connection as a feeling of,

> ... being part of something bigger than yourself, and not something that revolves around you ... more recognition that you're part of something else, you're not the most important part, you're not the only part. And everything sort of working together ... And not just connecting to animals and trees ... but being here with other people as well. I feel like you're more attuned to other people's needs and the way other people are feeling.

The understanding that Andrew articulates that humans are just another part of Country, fits well with the more-than-humanisms of contemporary cultural geography that he would have encountered through his university studies. There are close parallels between the ideas of post-humanism, particularly its decentring of the human subject (Braidotti 2019), and what Andrew understood to be the messages that Uncle Lex and the other Elders leading the camp were trying to communicate, such as listening to Country rather than imposing your will on it. These lessons were taught by the Uncles through day-to-day activities such as wood gathering. Andrew recounted how during one such expedition, a dead branch was suspended in a tree and could not be brought down, but "after a few ... seconds of pulling it and trying to dislodge ... Uncle ... was sort of like, 'It doesn't want to come down. Let's just leave it there. And maybe another day, it will come down'". By leaving the branch in the tree, the Elder was teaching about respecting the agency of Country, and modelling the spirit of "patience, humility and respect" Uncle Lex explained embodies Yanama Budyari Gumada, which in the Dharug language means "walking with good spirit" and is the theme of the Culture Camps (see Chapter 4).

The reciprocal relationship with Country Simon described was also one that Andrew understood the Elders were trying to communicate as a proper relationship between humans and Country,

> ... messages like ... there's also the importance of not just taking and taking, which is that sort of colonial law, mindset, where just take what you can, and screw everything else, where the importance of, if you are taking things, to use it all, to make sure you give back with things like ... replanting, or the activities that we've been doing, instead of just taking it all and leaving it all degraded, which is what has happened during the past.

This message is also in line with principles of sustainability, which form part of the discourse of natural resource management, identified in previous chapters as a recurring theme both in visitor interviews and in protected area interpretation (Howitt et al. 2013). While it could be regarded as a postcolonising discourse when it is prioritised over Aboriginal understandings of Country, ideas about sustainability also provided an entry point for non-Aboriginal people to understand the ongoing contemporary relevance of principles of custodianship as practiced by Aboriginal peoples.

A further Aboriginal ontology communicated to participants was that of relationality with the human and the more-than-human, an ontology that Aboriginal peoples continue to use to describe their identities and cultures (Dudgeon and Bray 2019; Tynan 2021). In fact, Kombumerri and Wakka Wakka philosopher Mary Graham, with Morgan Briggs, argues that "thoroughgoing attentiveness to relations, and thus to the patterns, ethical obligations, and contingencies that arise with those relations" are "the most persistent feature of Aboriginal ordering and philosophy" (Brigg and Graham 2020, n.p.).

Picking up on Andrew's point about being part of something bigger than oneself, Simon identified the message,

> that probably stood out the most for me, is the idea … of community and working together. And the importance of … leaving … your ego, and your selfishness and trying to move yourself away from that [by asking], 'How can I help?' … Like, you know, the community, this group?

"Bringing the human spirit to Country", UKTNP, May 2019

I'm in a gully looking up at a red sand dune, covered in spinifex and low-lying shrubs and trees. Some haven't survived the dry weather but even in death retain a delicate skeletal beauty.

Colours are grey/green, yellow/grey/green, the pale flax of dry spinifex, the orange/red of the earth and the astounding leaf green of some of the gums, roots finding water who knows where, and overall, the sky, an eternal blue.

Time passes strangely, neither fast nor slow. I don't feel it passing … Uluru is just over the sand dune, but I'm fascinated by small things, the fleeting, the unobserved. The withered shrub, feathered by afternoon light, the circular traces of spinifex in the sand left by wind, tiny gum nuts on their white stems. The dainty bush that wrens visit briefly.

Here they are, back again, and I'm recaught in the living moment. The crickets start to sing. I'm very still. My body vibrates with the evening bird song. The sound of my page turning crackles in the air.

Only one more evening here … I realise I've added sky blue to the green and orange paint marks on the old outside table (Figure 7.1).

Andrew describes a process in which he actively worked to transform his own thinking from a focus on his own individual wants and needs to those

Figure 7.1 Behind the sand dune, "Rangerville", UKTNP. Photograph by the author of painting by the author.

of the collective. This was in line with the message conveyed by Uncle Lex that Aboriginal social relations are more communally orientated than is the case in what Andrew described as "individualistic" "atomised" "capitalist" societies, in that "Indigenous people [usually] ... did things in a group, ... hunting in a group, you'd always be sort of sharing things as part of a group".

Participants were also taught to relate to Country in the present moment, including to other humans as part of Country, by being aware of what was happening around them. Cathy, another Yellomundee participant, found this message humbling because "... you're not saying, 'Oh, well, what am I going to do now?' It's like, 'Well, what's happening around me?'" Cathy thus developed a growing awareness that she existed, of necessity, always in relation to human and non-human worlds, and that she could turn her attention to these interconnections in a mindful way and perhaps engage with them more consciously. Andrew noted the sound of the water and bird calls and the feel of the dirt under his bare feet, while Simon was attuned to the sound of the river, the rain on his tent at night and the smell "of greenness in the air". Cathy and Sylvia also noticed "the smell of the river", and the sound of the rain. Camp participants' growing awareness of Country was necessarily embodied and sense-based, encompassing not only the normally dominant sense of sight but also that of sound, touch and smell, what Pink might term a "multisensory, emplaced experience" (Pink 2015, 96).

Andrew shared the insight he had gained from Uncle Lex, that "just even being on Country and sort of bringing the human spirit to Country without sort of even doing anything ... but just being here and sort of connecting here is still an activity". Several visitors to Uluru-Kata Tjuta National Park similarly emphasised that the park had to be experienced firsthand to be appreciated, rather than vicariously through books or other media. Frank, a retired visitor from NSW, interviewed while doing a guided base walk around Uluru, explained that the forms of interpretation he found most effective were "[p]artially the signage, partially the guide's monologue and partially both of those things while standing in the place, Not reading it in a book". He believed that the content of the interpretive signage could only be "fully understood within the place that it's written about". The importance of experiencing a place and hearing the accounts about that place firsthand, where they happened, was also emphasised by two senior Anangu men I consulted as the best way to understand Uluru-Kata Tjuta National Park.

There are synergies here with Bawaka Country et al.'s (2017) understanding that Country is sentient and agentic and has a role in co-constituting what is learnt. This was certainly my experience when I lived in ranger housing behind Uluru, close to the town of Mutitjulu while doing my fieldwork. The property had a collection of trees, like Blue Mallee, with gum nuts that attracted local birdlife. Most evenings I sat at the back of the house, nestled behind a sand dune, listening to the quiet as I watched the colour fade from the sky (Figure 7.1). Quiet times at Yellomundee, and the lack of a rigid structure to activities, although this was challenging for some of the participants due to the normally highly structured nature of their lives, also provided an opportunity to passively connect with Country.

The different way of living experienced at the Culture Camp prompted a new awareness of time for one participant. As Andrew explained,

> ... your concept of time changes. Because, I guess, when you're back in Sydney, you're always worrying about the time because it's like, well, 'Where've I got to be? ... What time is my bus? Is it running late?' Like you've sort of obsessed with time? And it's a very sort of structured concept of time, but then coming out here it's like, pretty much it could be any time of day, it doesn't really matter.

This capacity to simply be rather than do as they brought "the human spirit to Country" allowed space for the emergence of new ways of being, where time was experienced more as an unfettered flow than a regimentation.

Yellomundee participant Sylvia also felt that she had gained insight into a different way of viewing the world and thinking, especially "with the weeding, I'm not just like, 'This [plant] is good, this is bad', it's like, everything has a purpose, and there's a use and it's connected'". This could be regarded as both an Aboriginal approach to valuing Country and a post-human insight, as it is only humans who define plants as weeds on the basis that we have no use for them, or that they damage other plants that we value, such as the Casuarina underneath the balloon vines at Yellomundee. Sylvia's insight served to further de-centre the importance of the human in this landscape and was also closely related to what was promoted at the camp as the sustainability of Aboriginal lifeways, where "everything has a purpose" is connected and nothing is wasted.

I found this discourse of sustainability simultaneously appealing in the way it valued Aboriginal ontologies but also disquieting, because it seemed to romanticise Aboriginality, even though it was a form of self-representation by the Aboriginal Elders who offered the camp. Another disquieting but simultaneously seductive aspect of the Culture Camp from my perspective as a non-Aboriginal "participant observer" was the way that the Aboriginal ontology of Country was made available to non-Aboriginal camp participants (see Chapter 4). This was a deliberate strategy of the Traditional Custodians providing the camp, in line with the belief that everyone who lives on Country is part of it, and that we should "take seriously not just the need to care *for* Country, but to care *as* Country; to recognise that humans and Country exist together in a state of co-becoming" (emphasis added, Bawaka Country et al. 2013, 195). While I recognise the significance of these shared connections, it also seems important not to lose sight of the necessarily different relationships that Aboriginal and non-Aboriginal people have with Country. These differences arise not only in our disparate socio-political positionings and collective histories under Australian settler-colonialism but also in the responsibilities and intimate, identity-based connections that Aboriginal peoples have with ancestors, human and non-human kin, creation accounts and places of cultural significance.

Although the Culture Camp encouraged participants to build a relationship with Country, and the participants I spoke to believed they had begun to do so, I argue that, as non-Aboriginal people, there are inevitable limitations to the type of relationship that they could achieve. These limitations can be better understood by considering the meaning of Country to Dharug and Gundungurra Traditional Custodians I consulted as part of the research. A Dharug Traditional Custodian of the NSW Blue Mountains explained that for him, the idea of Country was "very encompassing, it's sky, the waters, people that have been, people that are coming … and Country for me is my area of responsibility", while a Wiradjuri Traditional Custodian and Blue Mountains cultural tour provider reflected that to him, "Country means everything. I've learnt how to love the space that I'm in and pay respects to the old fellas". Similarly, Gundungurra Elder, Aunty Leonie, who asked to be identified by name, explained, "I love to connect [to Country] because of my ancestors … We get the feeling because of our connection to this land, through our mum, and grandfather and grandmother. The connection".

Such meanings, although personal, resonate with Aileen Moreton-Robinson's (2022, 259) description of Indigenous origins as "in and of the earth", shared with and derived from "creators, ancestors and relatives that inextricably unite us with the earth and our respective shared territories". This sovereign relationship is linked to "origin stories that emanate from and connect us as humans and non-humans through relations and kin to all that Mother Earth and our creators made", such that "ways of being Indigenous are inextricably connected to being in and of our lands" (Moreton-Robinson 2022, 259). It seems clear, then, that as non-Indigenous people do not share the same custodianship responsibilities or connection to land, to origin stories and to human and non-human kin, they cannot share the same (or even an approximate) relationship with Country. This is not to say, however, that non-Indigenous people cannot come to grasp aspects of Indigenous ontologies and apply this learning in ways that are meaningful to them and that change or even transform them.

Visitor epiphanies and ongoing, sustained "transformations"

It may have been access to these Aboriginal ontologies, modelled by the Elders who led the camp, that resulted in Yellomundee Culture Camp participants articulating the most significant overall learning and change of any group of cultural tour participants that I consulted. The relatively long duration of the activity, the camping format that led to a sense of close connection to the landscape and participants' direct and ongoing access to the Traditional Custodians providing the experience, also contributed to these self-defined "transformations". This experience thus represented the type of "multi-sensory, complex social interaction" that Hughes and Morris-Saunders (2005, 161) identify as one advantage of the active guided tour format over passive forms of interpretation such as signage. For Andrew, who took part in a follow-up interview about his experience after four months, the shifts in thinking described above had not only been maintained but had developed further, resulting in resolutions about how he wished to apply these understandings in his future career.

After participating in the camp, Andrew reported that he immediately began reading more about Indigenous cultures, explaining that this reading "sort of sparked a change in addressing a lot of my own biases … all these new ideas, and all this new learning and I was like, 'I really want to get into that now'". Andrew also spoke about wanting to work on projects with Aboriginal peoples in his future profession of planning under their leadership, revealing how coming to see Aboriginal peoples as knowledgeable and authoritative leaders through cultural tour participation, as argued by Travesi (2018), can have broader applications. Other cultural tour participants, such as those involved in the Maraku Arts cultural experience (as part of a one-week tour led by a small group touring company), also reported dramatic insights and changes in behaviour. A follow-up interview with one such participant, Sue, a British backpacker in her early twenties, revealed that her increased understanding of A<u>n</u>angu and their cultural heritage had deepened to self-identified empathy, explaining "[w]e're both [her partner and herself] definitely changed, so, our opinions or, it's more a deeper understanding and now empathy for the Aboriginal people".

This "higher-level empathy" (Coplan and Goldie 2011, xxxiii-xxxiv) also led to changes in behaviour, including asserting a different viewpoint when Australian locals promoted culturally disrespectful practices such as climbing Ulu<u>r</u>u. As Sue recounted,

> ... we even speak to Australian local people, and they say 'Did you go climb the rock?' And we're like, 'No, we didn't, we didn't want to do that, the tour was very informative in telling us it's very disrespectful to the Aboriginal community, and all the reasons why you don't do that. And then some of the Australian locals will be like, 'Oh, now it's time to climb, ... we know it's closing soon if it hasn't closed already'. ... Sometimes their first question is 'Have you climbed the rock? And it's like 'No, it's not about that. That's not why we went to Ulu<u>r</u>u. We're there to learn about the Aboriginal people ... How the rock has spiritual significance'.

Not only did Sue and her partner, Thomas, become advocates for culturally respectful tourism, they also wanted to share what they learnt on the tour with others. In so doing, they modelled the type of socially engaged empathy that is carried through into action, which Smith and Campbell (2015) argue is necessary if the insights visitors to heritage places obtain are to translate into meaningful social change. Such outcomes provide hope for the Reconciliation Tourism discussed earlier in this chapter (Galliford 2010; Higgins-Desbiolles 2003). Sue's response also resonated with some of the positive changes reported by visitors to Bawaka Cultural Enterprises, who discussed "an increased willingness to share their perspectives and insights on Indigenous communities and worldview with others in social situations" (Lloyd et al. 2015, 14). A similar desire to share what was learnt, but with her family rather than strangers encountered while travelling, was articulated by Sylvia, a participant in the Yellomundee camp. She explained,

> I want to try and share that with my family, with my dad ... because honestly, he's very ignorant, I don't think he cares. My mum would be so interested in what we've done ... same as my sister, she's doing primary education ... I'd love to share what we've learnt here with her.

While participants in Aboriginal-led cultural experiences such as the Yellomundee Culture Camp identified the most profound learning and change, visitors exposed to other forms of Aboriginal cultural heritage interpretation also reported epiphanies that were potentially transformational. For example, Derrick, introduced in Chapter 6, had what he described as a complete change in perspective after experiencing the Tjukurpa tunnel at the Uluṟu Cultural Centre. As discussed in the previous chapter, Derrick fundamentally changed his viewpoint on the Uluṟu climb, which also prompted a critical appraisal of himself and other settler Australians. Although he travelled to Uluṟu-Kata Tjuṯa National Park to see Uluṟu and climb it, he explained that he abandoned this goal "after going through the Centre and reading and learning a bit". He claimed that coming to understand the Aṉangu perspective on the climb presented in the Tjukurpa tunnel, had "completely changed me in many ways in a matter of hours". Reflecting on this change in his thinking, he confided, "I think when I first came here, I was actually quite racist through ignorance and now I really want to learn ... and I want to sort of get their [Anangu's] side of the story".

Derrick's capacity to engage with the interpretation at the Uluṟu Culture Centre, both emotionally and cognitively, underpinned the transformation in outlook he described. Importantly, Derrick's changed thinking related not only to the Uluṟu climb but extended to a more generalised critique of colonial discourses of the "discovery" of Australia, demonstrating how visitor epiphanies of the moment can have broader implications. It was not only Derrick's capacity for affective and emotional engagement with the Aṉangu perspective presented in the Cultural Centre but also his ability to critique postcolonising paradigms of discovery, exploration and national achievement that facilitated his shifts in perspective. His transformation thus occurred in part through a process that Mezirow (1978, 104–5) identifies as "taking the perspective of others". Mezirow (1978, 104–5) explains that this "is not simply role taking for it also implies a conscious recognition of the differences between one's old viewpoint and the new one and a decision to appropriate the new perspective as being of more value". Because Derrick was able to articulate the ways his viewpoint had changed, his "perspective-taking" (Mezirow 1978, 104–5) was not simply affective but also an outcome of cognitive processes, involving conscious decision-making to reject his former way of thinking.

A repeated response shared by participants across interpretation types (apart from those interviewed at the base of the climb site) was a desire to learn more about the cultures of Aboriginal peoples following their Aboriginal cultural tourism encounter, including through continued engagement with them. It was as if their encounters in place had awakened their interest, representing only a first step in what would be an ongoing process of engagement and learning. Lloyd et al. (2015, 14) reported a similar outcome for participants in Bawaka Cultural Enterprises tours, who expressed "a greater interest in news stories and literature involving Indigenous issues and communities" following the tour.

The visitors to Uluṟu who learnt the least and reported fewer significant changes were those interviewed at the base of the climb after having climbed

Uluṟu. Only one of these six participants reported a changed perspective, and this was simply the banal appreciation of "how big [Uluṟu] is" (Angus, domestic Australian visitor), revealing that his understanding of the landscape had not moved beyond the superficiality of visual spectacle. Visitors who climbed generally arrived in Uluṟu-Kata Tjuṯa National Park with a single intention, and this did not extend to engaging with the Aboriginal cultural heritage of the park. The different responses from visitors at the climb site are also indicative of a possible selection bias in the research arising from recruiting just over half of the Uluṟu sample from visitors who chose to participate in cultural tours and visit the Uluṟu Cultural Centre. These visitors were possibly more open to engaging with Indigenous cultural tourism than those selected randomly at the climb site.

Limitations and barriers to "transformation"

This section focuses on some of the barriers to change experienced by visitors who engaged with Aboriginal cultural heritage in protected areas and the implications of this for the design of interpretive content. While many of these limitations arose from the entrenched nature of a settler-colonial mindset among visitors, they also appeared to be related to the discursive silences of some interpretation, the format of tours and the personal circumstances of some visitors. Although participants in Traditional Custodian led cultural tours experienced the most numerous and significant shifts in understanding and behaviour, not all of this group shared the same level of engagement with the landscape or with tour content.

For example, Jane, first mentioned in Chapter 4, summed up her emotional response to the Patji cultural tour as feeling, "I'm outside of all this. I respect what's being said, but I can't really be part of it". Unlike the feeling of connection arising from visitors' experiences of other cultural tours, such as those offered by Bawaka Cultural Enterprises, which involved longer stays and closer involvement with the life of the community (Bawaka Country et al. 2017; Lloyd et al. 2015), Jane's one-day cultural tour did not engender a strong feeling of connection with Country or A<u>n</u>angu culture. I thought at the time that this response might have been partly a result of the physical separation from the landscape that resulted from being on a bus, rather than participating in a tour with the type of embodied connection to place engendered by walking or camping. However, these options would have been too challenging for travellers like Jane and her partner Arthur, due to health issues associated with ageing.

The underlying reasons for Jane's feelings of separateness were revealed in a follow-up interview four months after the tour. She confided that personal trauma arising from a bereavement and an adverse medical diagnosis received by her partner, Arthur, meant that they could not really take in what was happening on the tour. Jane's feeling that she was "outside all of this" and "can't really be part of it" conveys a strong sense of feeling "other" and of not being physically present that one can feel after a shock or trauma. Jane's containment was at the opposite end of the experiential spectrum to the embodied relationality discussed by all of the Yellomundee participants. Relatedly, Walkerdine, drawing on the work of Guattari to inform her own work on the spatial identities of working-class women in London in the early 1990s, argues that,

> ... change (and development...) is about being able to move beyond a personal world of constraints, limits, and coordinates (the existential territories) to universes in which those things disappear and one is carried along on something new, one is able to open oneself up to something new 'I am no longer what I was before, I am swept away by a becoming other, carried beyond my familiar existential territories'.
> (Guattari, 1995, 93, cited in Walkerdine 2013, 759)

The feeling of separateness Jane articulated meant that although she felt "other", she was not, as it were, "swept away" by this feeling in a way that transformed her, unlike Derrick's encounter with the Uluru Cultural Centre, whom this description aptly fits. This suggests that regardless of tour format, medium or content, the varying personal circumstances of participants and the way in which these affect their physical and mental well-being, including their capacity to be mindful of what is happening around them (Ling Tan et al. 2020), inevitably informs their experience. These subjective factors, while not necessarily imposing a limitation, demonstrate the acuity of Witcomb's observation (2015a, 138) that it is not simply a matter of curators or interpreters imposing meaning on passive visitors but that "the subjectivity of the visitor and their interaction with these components" is actively involved in co-constituting meaning.

Limitations to visitors' learning and change could also arise from the tendency of cultural tours such as the Blue Mountains Walkabout Tour to present a stereotypical and monolithic depiction of Aboriginal peoples and cultures. As evident from the Blue Mountains Walkabout guide's comments in Chapters 4 and 5, he assumed that prior to colonisation Aboriginal peoples lived in an unmitigated state of happiness and wellbeing, evoking Rousseau's discourse of the "noble savage" who supposedly existed in blissful harmony with their environment (Dodson 2003; Hinkson and Fullenwieder 2019; Langton 1996). Accordingly, participants like Katrine believed that the tour had conveyed that Aboriginality,

> ... comes with rituals and being in the middle of nature and so on. And we've lost track of it because we live in cities and our lifestyle has completely changed. And by telling all the stories today and showing us how they were told, he kind of reconnected us with what it would have been before we moved to cities.

Rather than recognising that Aboriginal peoples have complex and changing identities and that they now primarily live in urban areas, Aboriginal peoples were perceived as identical to other Indigenous peoples, who were presented as not only uniform but pre-modern. The discourses of participants in the Blue Mountains Walkabout Tour thus appeared to reproduce the simplistic representations of Aboriginal peoples present within tour content.

Some visitors to Uluru-Kata Tjuta National Park also developed more entrenched stereotypes of Aboriginal peoples following their cultural tour participation, coming close to the type of dangerous binaries Bawaka Country et al. (2017) warned about earlier in this chapter. Retiree Jason, discussed in previous chapters, came away from the Maraku Arts cultural experience with a dualistic perception of Aboriginal peoples based on a model of authenticities and deficits, which solidified over time. In a follow-up interview four months after the tour, Jason referred to remote-living Aboriginal people such as Anangu as "the 'real Aborigine'" as opposed to Aboriginal peoples living in urban areas whom he regarded as "poisoned by white civilisation: the alcohol, the drugs, petrol" (see Chapter 5). When Jason said in the initial interview that the tour had increased his knowledge and respect for Aboriginal cultures, he was referring to Anangu living close to Uluru, rather than urban Aboriginal peoples. Changed understandings articulated by visitors were thus not necessarily for the better or could be partial, suggesting that the researcher may need to drill down to understand what is meant by these assertions.

Similarly, Cathy, a Yellomundee camp participant, retained certain unacknowledged colonial paternalisms at the end of the Culture Camp. Discussing her objection to media stereotypes of Aboriginal peoples, she argued that "the media … contribute to giving *our* Indigenous people a bad name" (emphasis added). Cathy was making the point that it is not unusual for popular media discourses to denigrate Aboriginal peoples and claim that Aboriginal culture no longer exists in the Sydney region. However, she used the possessive pronoun "our" to refer to Indigenous peoples, signifying a sense of ownership or proprietorship, reflecting the unconscious white paternalism that continues to permeate settler thinking (Moreton-Robinson 2015).

It is possible that a focus on more-than-humanisms and popularised Aboriginal ontologies, while giving Yellomundee camp participants insight into Aboriginal ways of being, failed to enlighten them about racialised discursive formations that continue to reproduce power differentials in the contemporary Australian context. As discussed at length in Chapter 5, "othering" discourses of Aboriginalism and white paternalism are deeply embedded within the settler psyche, informing the unquestioned assumptions about identity, land and nationalism that led some domestic visitors to Uluru-Kata Tjuta National Park to proudly describe it as "our country", and to identify the Aboriginal heritage of the park as belonging to all Australians. It is important that cultural tours and other forms of interpretation of Aboriginal cultural heritage in protected areas work to disrupt racialised discursive formations and "settler structures of feeling" (Rifkin 2011, 350) by challenging visitors' stereotypes and assumptions about Aboriginal peoples and places. As Galliford (2010) notes, one way to achieve such a disruption is to include political and social content within cultural tours that reflect the contemporary lives of Aboriginal peoples. The alternative is to continue to reproduce narrow representations of Indigenous identities, as Bunten (2018) has argued, through a singular focus on traditional lifeways, rather than communicating the diversity of contemporary Aboriginal identities, including the ways that Aboriginal peoples relate to Country now.

Conclusion

My research suggests that participation in Aboriginal cultural tourism, most notably cultural tours led by Aboriginal peoples, could lead to profound changes in some visitors' understandings of Aboriginal ontologies and epistemologies and the way these are linked to places of importance to Aboriginal peoples. Not only did such engagement result in perspectival shifts as visitors began to "see" Country, but for some, prompted a revaluing of Aboriginal ways of being and an increased respect for Aboriginal peoples and cultures that have been devalued by settlers since colonisation. For some visitors, it was not a huge leap to move from an appreciation of the sophistication and sustainability of traditional Aboriginal lifeways to a critical appraisal of the invasion and colonisation of Australia as having profoundly negative consequences for Aboriginal and Torres Strait Islander peoples. The inclusion of Aboriginal perspectives on post-contact history in some cultural tours, foregrounding events such as the frontier wars and Stolen Generations, made the ongoing nature of colonisation apparent to some visitors. Indigenous cultural tourism was thus most powerfully transformative when visitors were able to generalise from a specific encounter or experience to critically reflect on Australian settler-colonialism as a dominant ongoing social and political formation.

However, as various commentators have argued (Bawaka Country et al. 2017; Galliford 2010), it is unrealistic to expect that visitors could be completely transformed by a single cultural encounter, particularly those which reproduced rather than challenged problematic discourses about Aboriginal "authenticity". My research suggests that the transformations experienced by some visitors were, at best, partial and incomplete, as they remained entangled within settler-colonial "mindsets" (Mezirow 2000, 7–8). It was even possible for commitment to such mindsets to be further entrenched by the visitor experience. Additionally, cultural tours could themselves reinforce pervasive stereotypes and misconceptions about Aboriginal peoples. This occurred because of the assumptions about Aboriginal cultures that permeated some encounters, including that Aboriginality is monolithic and unchanging. Unless visitors also develop an understanding of racialised discursive formations prevalent in the Australian context and can critically engage with stereotypes and misconceptions about Aboriginal peoples, it seems likely that the "transformations" they experience will necessarily be limited.

One way to encourage such perspectival shifts would be to provide opportunities for visitors to integrate what they have learnt through structured reflection and discussion, as recommended by Lloyd et al. (2015), although none of the cultural experiences I participated in offered this form of engagement. This points to the importance of cultural tour experiences that not only critically engage the visitor but allow opportunities for dialogue with other visitors and Aboriginal tour providers that may, directly or indirectly, address racialised stereotypes. Such experiences increase awareness, and because they engender "cognitive and behavioural responses" (Walker and Moscardo 2016, 1243), may have a more lasting impact on visitors. My research thus supports the argument that Aboriginal led cultural tours can contribute to reconciliation by decolonising the habitual modes of thinking of settler visitors, but that this is facilitated by tour content that encourages critical engagement in visitors.

Simply learning to revalue Aboriginal lifeways for attributes such as sustainability and relationality does not in itself constitute a form of critical engagement. This is because these concepts are based on an authenticity discourse of Aboriginality as essentially closer to nature, failing to account for the complex contemporary identities of Aboriginal and Torres Strait Islander peoples (Paradies 2006). Such revaluing may also extend to attempting to appropriate Aboriginal identities, such as a relationship with Country, which is both a cornerstone of Aboriginal selfhood and sovereignty, and in this way securing, albeit fictitiously, a settler claim to Australia (Wolfe 1999, 2006). These entanglements call to mind Braidotti's (2019, 42) assessment, that "as embedded and embodied, relational and affective subjects, we are immanent to the very conditions we are trying to change". As a consequence, transformation presents a significant but hopefully not insurmountable ontological challenge for the settler subject.

...

Postscript

This chapter has focused on the "transformations" of visitors to protected areas rather than the socio-cultural landscapes of the research locales. There can thus be a tendency to view such landscapes as static surfaces that visitor "transformations" occur on, rather than as living entities that also change, or "co-become" with visitors. The dynamic nature of the socio-cultural landscape of Uluṟu-Kata Tjuṯa National Park was brought home to me when I returned to the so-called "timeless land" promoted in tourist advertising two and a half years after conducting my park-based fieldwork. The purpose of the trip was to present a very abridged version of my findings to the park's Tourism Consultative Committee, a sub-committee of the park's joint

Aṉangu and Parks Australia Board of Management. I also found that this journey afforded me the opportunity to reconnect with the landscape of the park, one I found much altered.

I had been intending to share with the Tourism Consultative Committee details of visitor support for the climb closure and the strong objection to the climb voiced by a number of research participants. However, I was asked by the new Park Manager not to mention the former climb at all, as I was told Aṉangu "have moved on from that". This moving on was also apparent at the site of the former climb, where the signage asking visitors not to climb in a range of languages had been removed, replaced by an unobtrusive notice that simply read "Permanent Closure, 26 October 2019". I breathed a sigh of relief that the chain and posts that visitors held to make the ascent had also been removed, but most of all that climbers were no longer visually present in the landscape.

Other changes to the park were the result of the collapse of national and international tourism to Uluṟu due to the COVID-19 pandemic. This collapse was prompted not only by "fortress Australia's" closure to the international community but the closure of state borders to domestic travellers and the introduction of restrictive quarantine requirements. The need to ensure Aboriginal and Torres Strait Islander communities were vaccinated prior to reopening was also an important reason for this, although vaccine coverage in many such communities remains patchy. COVID had also reached the Aṉangu community of Mutitjulu within Uluṟu-Kata Tjuṯa National Park, including people I was acquainted with from my former fieldwork. It was impossible not to notice the lack of tourists in the park during my 2022 visit, as well as the eagerness of hospitality staff for news of the arrival of domestic flights, as if these were heralded events.

Returning to the Mutitjulu waterhole, I spent some time sitting on the shaded seat near the carpark, reluctant to venture into the hot February sun. I felt an absolute sense of peace without the presence of other visitors – it seemed possible to deeply connect with the place. The park was also much changed after the heavy rains of early 2022: everything seemed greener and there were many more birds than I'd noticed before. It was wonderful to see flocks of Zebra finches, wild Budgerigars and green parrots feeding young in the hollow of a dead tree, close to the walk to the waterhole. There also seemed to be more reptiles in the park, including the large Perentie that had reclaimed a rock at the base of Uluṟu. A mass of tadpoles writhed in the remaining water at Kantju Gorge and the Mutitjulu waterhole, a sign that the burrowing frogs had emerged to breed after heavy, prolonged rain. The Bloodwoods around Uluṟu were in flower. And the Mutitjulu waterhole had also been reclaimed: by Aṉangu.

When I visited the park in 2019 to do my fieldwork, I interviewed a young Aṉangu man who was the host of the Patji tour. He talked about wanting the Uluṟu climb to close so he could swim in the Mutitjulu waterhole again, like he did as a boy, before it was contaminated by visitors using the top of Uluṟu as a toilet and garbage dump. Two and a half years later, I arrived at the waterhole car park at the same time as two cars of Aboriginal men and boys, a few wearing board shorts. I knew they'd come to swim, as they confirmed during our brief conversation at the waterhole.

The following day I was back looking for a lost camera lens cap when another car arrived, this time containing a large Anangu family, also heading for the waterhole. I stayed back on the seat at the car park, not wanting to intrude on their time at the waterhole, wondering if Anangu felt a similar inhibition accessing it when the waterhole was overtaken by busloads of visitors, so much so that the new busload had to wait for the previous one to exit before they could enter. There's an obvious park sign targeting visitors at the waterhole that prohibits swimming, one clearly disregarded by Anangu, but perhaps not while visitors are present? I could not help wondering what would happen when mass tourism returned to Uluru, and whether Anangu would have the same access to the waterhole that they did in its absence? Or would the park again transform into a landscape dominated by visitors, recreating it once more visibly, but perhaps only superficially, given Anangu's deep connection with this landscape, as settler-colonial space?

Note

1 Although all cultural tour interview participants were offered the opportunity to participate in a follow-up interview on the longer-term impact of their cultural tour experience upon them, only four agreed to do so. Three of these were cultural tour participants from Uluru-Kata Tjuta National Park and one was a Yellomundee Culture Camp participant.

8 Conclusion

Heritage, tourism and incommensurable Aboriginal sovereignties

Endings and immersions

When I began this research project, I did not know that it would end with A<u>n</u>angu swimming in the Mutitjulu waterhole after the closure of the Ulu<u>r</u>u climb and heavy rain returned the water to a cleaner state. Nor could I have envisaged that the advent of COVID-19 and the almost total absence of tourists in Ulu<u>r</u>u-Kata Tju<u>t</u>a National Park when I returned in February 2022 would provide A<u>n</u>angu with the opportunity to reclaim the waterhole for their own use. Similarly, I could not have predicted that I would also find myself immersed in, of all places, the murky Nepean River during the Yellomundee Culture Camp, nor that my own relationship with the Blue Mountains and its many examples of colonial architecture would be drastically altered. A research project can thus be a journey of many twists and turns in which one not only engages in new experiences and encounters unforeseen events but arrives at conclusions that cannot be predicted.

My initial aim when commencing the research was to explore the ways that visitors respond to the interpretation of Aboriginal cultural heritage in Australian protected areas, but it was certainly not clear to me at the outset the significant role that settler-colonialism would play in this. However, once I began to engage with visitors to protected areas, it was soon apparent that Aboriginal cultural heritage is experienced in the socio-cultural context of Australian settler-colonialism. Although I had hypothesised that socially constructed meanings attributed to Aboriginality might influence visitor responses to interpretation within the landscape, I did not know what form these discourses would take, nor did I understand their implications. The research I have undertaken in Ulu<u>r</u>u-Kata Tju<u>t</u>a National Park, the Blue Mountains National Park, the Jenolan Caves and Yellomundee Regional Park has enabled me to explore such questions. It has also enabled me to assess the role that exposure to different interpretive media, including interpretive signage, digital forms of interpretation and Traditional Custodian led cultural tours has played in influencing visitor responses.

In recognition that visitors are embodied beings in direct engagement with Aboriginal heritage places, I have also examined if affect, emotion and

DOI: 10.4324/9781032654973-8

embodiment, in addition to discourse, play a role in visitor responses to interpretation in the landscape. The relationship that visitors were able to develop with Country, in fact, the extent to which they could access this Aboriginal ontology, and the implications of this for the politics of Indigenisation, was also considered. The extent to which visitors were changed by their encounters with Aboriginal cultural heritage in place, even "transformed" by them, was a further major focus of inquiry.

Research findings, contribution and implications

For a minority of visitors, engagement with Aboriginal protected area tourism produced life-changing epiphanies, such as a revelatory recognition of their own racism. While others reported no change, for most, the alterations in their attitudes, understandings and behaviour, although significant, were less dramatic. These included articulating an increased respect for Aboriginal peoples and their cultures and coming to understand Australian landscapes, including those close to the urban interface such as Yellomundee Regional Park, as Aboriginal Country. For some visitors, this involved entering into a different way of seeing, as they began to perceive Country as agentic and inscribed with meaning by creator beings through the Dreaming, coming to share, to a limited degree, in Aboriginal ways of seeing Country (Rose 2003). Longer-term changes were also identified, such as a desire to build on their increased knowledge and understanding through continued learning and engagement with Aboriginal peoples, and an increased willingness to advocate for respectful interactions with Aboriginal peoples in a tourism context.

As a consequence of recognising peri-urban landscapes as Aboriginal Country, some visitors came to perceive the totality of Australia as unceded Aboriginal land, troubling their own relationship with a place they understood as home. Others openly discussed the historic dispossession of Aboriginal peoples and called for meaningful redress of past and present injustices. In emphasising the transformative potential of visitors' engagement with places of Aboriginal cultural heritage, the findings of my research contrast with those of Smith (2014), that visitor engagement with heritage places was more likely to affirm visitors' existing views rather than change them. The meaningful "transformations" experienced by multiple visitors I spoke to suggest that engagement in Aboriginal cultural tourism has the potential to progress an agenda that goes beyond symbolic reconciliation in the contemporary Australian context. My findings in this regard build on the theoretical foundations of Higgins-Desbiolles (2003), who argued that tourism can play a role in reconciliation, ideas also taken up by Galliford (2010, 2012). They reinforce the findings of Bawaka Country et al. (2017) that engagement with Indigenous cultural tourism can be positively transformational for visitors, building on this work by discussing some visitors' willingness to engage with post-contact histories of dispossession and violence in the cultural tourism context (see also Lloyd et al. 2015).

The importance of the altered thinking, attitudes and behaviour of many visitors notwithstanding, not all visitor interviewees were positively changed by their cultural tourism experiences. Some continued to engage in forms of discourse that applied a classificatory settler logic to Aboriginal peoples, based on an unstated assumption that whiteness was a superior way of being. I highlight how this sometimes manifested as a white paternalism, evident in visitors' usurpation of the right to judge and assess Aboriginal peoples. Ironically, these assumptions of superiority were sometimes disguised beneath the façade of a "virtuous anxiety" (Slater 2019, 3), expressed as a desire to help Aboriginal peoples "progress" and to address the negative aftereffects of white colonisation. By linking white, racialised modes of thinking about Aboriginal people with visitors' engagement in Indigenous cultural tourism, my research contributes to an emerging literature on whiteness in the tourism context (Bandyopadhyay and Patil 2017; Erickson 2018; Stinson et al. 2021). It also harks-back to earlier work in the field of heritage studies which argued that heritage discourses can establish whiteness as normative, marginalising the heritage of those viewed as "other" (Byrne 2003b; Hall 1999).

While systems of classification and surveillance have been identified in the historic and contemporary *social* policies of Australian governments towards Aboriginal peoples (Armitage 1995; Dodson 2003; Klein 2020; Moran 2005), my research reveals how such systems are also implicated in protected area heritage management, interpretation and tourism. Foremost among these classificatory discourses was the tendency of some visitors to discursively construct Aboriginal peoples as "authentic" and others as "inauthentic". Markers of supposed authenticity included physiological differences from a white "norm", maintenance of tradition and language, remoteness and time-depth. Although an "authentic" identity was viewed positively by some visitors as a form of alterity that acted as a drawcard for cultural tourism, signals of inauthenticity were believed to include the moral decline that sometimes these very same visitors attributed to supposed violence and alcohol misuse among urbanised Aboriginal peoples. Through this research, I have demonstrated that forms of what I refer to as an "authenticity discourse" are not only present within visitor narratives but within the interpretation of Aboriginal cultural heritage present within protected areas.

Although other scholars have already identified stereotypes about Aboriginality within protected area interpretive signage (Clarke and Waterton 2015; Staiff et al. 2002), my research connects such representations to a pervasive authenticity discourse tied to settler-colonialism. In so doing, I engage literatures in the field of decolonial, settler-colonial and critical Indigenous studies not typically discussed in relation to heritage interpretation, demonstrating the relevance of these literatures to this field (Dodson 2003; Mignolo 2011; Moreton-Robinson 2015; Rifkin 2011; Slater 2019; Wolfe 1999). It is important, however, to acknowledge the early work of critical heritage scholars such as Byrne and Harrison (2004) in this regard. Byrne (1991, 1996, 2003a) linked heritage with colonialism in the settler context, arguing that the language of heritage has been used to appropriate the material culture of Aboriginal peoples within the landscape to claim a "deep past" for settler Australians as a form of nation building. My work links the dated but still extant heritage interpretation at Red Hands Cave in the Blue Mountains National Park with the heritage discourse Byrne identified, suggesting that his analysis still has traction. By discussing the more recently developed interpretation in Uluṟu-Kata Tjuṯa National Park and on the Jenolan Caves mobile phone app and guided tours such as Patji, led by A<u>n</u>angu, I seek to contribute to these debates by identifying specific ways in which these more recent offerings decolonise protected area heritage interpretation.

While interpretation in jointly managed parks such as Uluṟu-Kata Tjuṯa does now foreground and prioritise A<u>n</u>angu ontologies such as Tjukurpa, and is in this sense decolonising, on a deeper level it continues to reflect ongoing colonial ontologies such as a nature/culture dualism. It does this by classifying Aboriginal Country as a cultural landscape to be conserved via natural resource management and expressing Tjukurpa as cultural knowledge that contrasts with Western science. Although the critique of a nature/culture dualism in protected area management and interpretation is not new (Brown 2021; Byrne et al. 2013; Harrison 2015; Bird Rose 2003; Bird Rose et al. 2003), its persistence in my research locales suggests that this dualism is deeply embedded in the thinking of protected area managing agencies. It is possible that utilisation of terms such as "Traditional Custodian" or "Traditional Owner" in protected area management discourse in association with a pre-existing nature serves to perpetuate a nature/culture dualism.

While terminology such as "Traditional Custodian" or "Traditional Owner" is typically used to acknowledge the ongoing relationships that Aboriginal people have with certain places, such terms may be implicated in an authenticity discourse, as they imply that peoples thus described have intrinsic ties with tradition, however, conceived. These terms may thus inadequately convey the dynamic and changing nature of Aboriginal people's relationship with Country. They may also contribute to visitor expectations about how those thus described should behaviour and what they should know, delimiting and constraining the identities of A<u>n</u>angu and other Aboriginal peoples. Such terms have become pervasive in the Australian context and have also to some extent been adopted by Aboriginal peoples, including the Dharug people I interviewed, and form part of their self-identity. Although these terms may be linked with an authenticity discourse about Aboriginality, they also have value and utility to some Aboriginal peoples. As a Dharug interviewee explained, the term "custodian" is preferred over that of "owner" because Aboriginal people do not see themselves as owning Country, but rather as caring for it. This term is thus closer to the A<u>n</u>angu term of *nguraritja*, which translates as "someone that belongs to a place" (Australian Government 2020, 2). Belonging to a place and all that this implies is not something of the past, even though the configuration and use of places may have changed, for example, through tourism or urbanisation.

Arguments that authenticity discourses linking Aboriginal peoples with tradition are present within tourism, protected area management, heritage discourse and archaeological practice are not new. Scholars such as Byrne (1991, 2003a) and Harrison (2004) argued that the work of archaeologists prior to the new millennium, in its focus on Aboriginal pre-history, tended to obscure the importance of post-contact places, both tangible and intangible, to Aboriginal peoples. This, they argued, contributed to an authenticity discourse that labelled Aboriginal peoples with clear ties to pre-colonial cultural practices as authentic, and other Aboriginal peoples, mainly those in south-eastern Australia, as inauthentic. This misapprehension carried over into the neglect of post-contact Aboriginal history within the heritage system, in favour of a focus on settler history. Although this trend is now changing, evident in the formal heritage recognition of Aboriginal people's attachment to places in the post-contact landscape such as former missions and cemeteries, my research identifies a continuing tendency of protected area agencies and the heritage consultants contracted by them, to neglect the co-existing histories of Aboriginal and non-Aboriginal peoples. This neglect is apparent within heritage interpretation in the landscape, and the outcome is to reproduce an authenticity discourse about Aboriginal people in protected area settings.

Utilising the ideas of Foucault and theorists who apply his ideas to discourse analysis (Jäger and Maier 2016), my research builds on the earlier work of scholars such as Byrne and Harrison, to identify this authenticity discourse as a discursive formation. This conclusion is based on the recurring presence of an authenticity discourse across a number of discourse planes, including visitors' spoken discourse in an Indigenous cultural tourism context and protected area interpretation. Not only does this discourse appear across a number of planes, it exhibits characteristics such as "regularity" (Foucault 2002, 41) and a "systematicity of ... ideas" (Mills 1997, 15) which are features of discursive formations. While Fforde et al. (2013) have described deficit discourses about Aboriginality in similar terms, my research reveals how romanticised ideas of Aboriginal authenticity, while positively reframing Aboriginality, constitute a powerful discursive structure that engages with these deficit discourses to constrain Aboriginal identities to one of either caricature. However, this is not to say that Aboriginal and Torres Strait Islander peoples do not have agency to resist these constraints, and to reframe their identities on their own terms, often on the basis of their own enduring ontologies and epistemologies.

In considering this issue, it is significant as to whether these representations of "authentic" Aboriginalities are a form of self-representation *by* Aboriginal peoples, or representations *of* them by protected area managing agencies, for example (Clifford 2013; Spivak 1990). As Clifford (2013, 14) argues, "... if the essences and traditions invoked by indigenous (sic) activists sometimes seem to repeat older colonial primitivisms, ... they do so at another moment and for new purposes". Clifford (2013, 16) explains that such performances of identity not only reflect "deep belief" but may also represent "the tactical presentation of self". The Aboriginal peoples who offered the Yellomundee Cultural Camp, for example, shared revived cultural practices such as Smoking Ceremony, dance and use of ochre with visitors, as a form of political assertion of their survival as Aboriginal peoples in the face of a colonial regime that, in the words of one Traditional Custodian, had attempted to destroy them. Reclamation of such cultural practices at the Yellomundee Culture Camp was not about asserting an "authentic" unchanging Aboriginality, but a strategic act of political resistance and a demonstration that important cultural practices are continuing in south-eastern Australia, despite colonisation. Traditional practices were thus reinterpreted in a way that was both self-defined and served the present moment. My research in this area thus reinforces existing arguments that observe that strategic essentialism has utility for Indigenous peoples, including within a cultural tourism context.

Traditional beliefs and practices are also a continuing aspect of culture for many A<u>n</u>angu, including those I interviewed in Ulu<u>r</u>u-Kata Tju<u>t</u>a National Park, and it is reasonable to expect that they would wish to share these beliefs and practices with visitors. Such beliefs include the three main creation accounts associated with Ulu<u>r</u>u that visitors are permitted to know. However, my research suggests that what is missing, in Ulu<u>r</u>u-based on-site signage and in some communications with visitors through cultural tours, is an indication of how such practices and beliefs have adapted and changed over time. This may arise from a lack of awareness that it is important for the cultural tourism experience to communicate representations of Aboriginal peoples as peoples who adapt and change, in order to challenge cultural stereotypes. The continuing presence of an authenticity discourse in protected area interpretation may also derive from perceived consumer demand by visitors to engage with "authentic" Aboriginalities.

Attempts to cater to perceived visitor expectations for an "authentic" experience were apparent in web and print based promotional material of Aboriginal-led cultural tours developed for the tourist market, discussed in Chapter 4. While these did indeed reflect the expectations of some visitors, other domestic and international visitors to both the NT and Blue Mountains research locales, were critical of stereotypes of authenticity, keen to know how Aboriginal peoples live now and to learn about the post-contact histories of protected areas. These were subjects some visitors reflected were missing from the interpretive signage in Ulu<u>r</u>u-Kata Tju<u>t</u>a National Park, and products such as the ranger guided walk, also offered in this park. Interviews with non-Indigenous protected area managers suggest that the reason for this absence from state-sponsored interpretation is the continued belief, after Tilden (2007), that the primary purpose of interpretation in protected areas is to educate visitors on pro-environmental messages, and that everything else is subsidiary to this. International visitors who participated in the Blue Mountains Walkabout Tour expressed a similar willingness to learn more about the post-contact history of the Aboriginal peoples of this region, content that was largely absent from this tour. My research thus highlights that the assumption that all visitors are searching for a "primordialist" (Theodossopoulos 2013b, 398–9) authenticity is misplaced, and that some visitors are eager to learn of the contemporary lifeways of Aboriginal peoples.

However, including site-based interpretation of post-contact history such as the 1985 "hand-back" of Uluru-Kata Tjuta National Park to Anangu at the place it occurred, requires the capacity to link the cultural, social and historic with the landscape of the park, a holistic way of thinking that may be difficult for protected area managing agencies to grasp. It also requires a willingness to be open and honest about an often fraught post-contact history, that included violence against Aboriginal peoples and other injustices, at World Heritage sites accessed by international visitors. The insights offered by this research, including that many visitors I spoke to were open to potentially confronting interpretation of this type, provides a strong rationale for protected area management authorities to address the silences in their own discourses and management practices that reproduce settler-colonialism.

Such silences were challenged in Anangu controlled cultural experiences such as the Patji tour on Anangu, Pitjantjatjara, Yankunytjatjara (APY) lands. This tour provided insights into the post-contact histories of Aboriginal peoples, including touching on relatively recent events such as Anangu's land rights struggle to reclaim Uluru-Kata Tjuta National Park and the Uluru family's own experience of the Stolen Generations. Interpretive content of this kind historicised Anangu by presenting them as peoples with a contemporary history that has impacted their identities, not just a deep past. The tour is promoted to visitors as covering subjects such as Anangu's land rights struggle to regain the land that is now the park, and their willingness to participate in it again suggests their openness to interpretative content of this type. Aboriginal cultural tour providers have the capacity to not only use cultural tourism to promote an agenda of reconciliation but as a vehicle for decolonisation, based on the type of narratives they employ and the experiences they offer. As revealed in Chapter 7, cultural tourism also has the potential to shift visitors' ways of seeing Country, so that they come to understand it as an Aboriginal place rather than simply as an aesthetic landscape. By using ethnographic fieldwork to reveal how some cultural tours promoted complex, contemporary understandings of Aboriginal identities, I hope to provide those responsible for developing interpretive experiences of Aboriginal cultural heritage with insight into promoting multi-layered and contemporary representations of Aboriginal peoples.

The interpretive model that appeared most effective in engaging and "transforming" the visitor was immersive, long-term cultural encounters on Country, such as the week-long Yellomundee Culture Camp. This experience seemed to be most impactful because of the prolonged opportunity it afforded for engagement with the Aboriginal peoples conducting the camp. Although other research such as that by Bawaka Country et al. (2017) also identified the transformative potential of visitors' spending an extended time on Country with Traditional Custodians, my research identifies its value in contrast with shorter Aboriginal-led cultural experiences. The Yellomundee Culture Camp provided participants with the opportunity to camp on Country and to engage with Country by weeding, gathering firewood and swimming in the river, or to simply "bring the human spirit to Country" by being rather than doing, as one participant described it. This experience led some participants to feel that they had begun to establish a relationship not only with the Traditional Custodians who conducted the Camp but with Country itself. The opportunity for continuing engagement with the Traditional Custodians who offered the Yellomundee Culture Camp was also welcomed by participants, although the outcomes of further engagement after the culture camp/tour had concluded were beyond the scope of my research.

Other cultural tourism experiences, such as exposure to the Uluṟu Cultural Centre and the interpretive signage at the base of Uluṟu, affected visitors in a range of ways, from mildly engaging their interest to profoundly changing their perspectives both on themselves and the Australian nation as a whole. Encountering this interpretation also led some visitors to reconsider or abandon culturally offensive behaviours such as the Uluṟu climb that they had formerly been committed to. These visitor responses demonstrate that not only immersive cultural experiences, but on-site interpretive signage, museum-like displays such as those found at the Uluṟu Cultural Centre, and digital interpretation using mobile phone apps, could be extremely meaningful, indeed transformative, for the visitor.

Effective interpretive examples had in common that they were either developed in consultation with Aboriginal peoples or produced by them, unlike the signage at Red Hands Cave, which excluded a Dharug and Gundungurra voice and perspective. Consequently, I conclude that it is the involvement of Traditional Custodians in developing interpretive material, and their capacity to authorise or approve it, that is the decisive factor in the success of the interpretation in meaningfully engaging the visitor, rather than the interpretive medium itself. My research thus makes the important link between the effectiveness of interpretation, including its capacity to engage in truth-speaking, and Aboriginal control over content and delivery and hence with Aboriginal self-determination and sovereignty. In contrast, the silences within state-sanctioned interpretation on colonial history represent a discursive prohibition (Foucault 2002, 1981) that exercises power by limiting what can be said about a subject. The failure of the October 2023 Referendum on a representative Aboriginal and Torres Strait Islander Voice to the Australian Parliament, makes the question of truth-speaking about Australian history in all fields of Australian life, including protected area interpretation, all the more urgent.

Stark differences between interpretation developed after consulting Aboriginal peoples and interpretation installed without consultation with them were apparent when comparing interpretive signage within Uluṟu-Kata Tjuṯa National Park with the interpretation associated with Red Hands Cave in the Blue Mountains National Park. Such differences not only relate to the era in which this interpretation was installed but can be traced back to the disparate governance arrangements and formal ownership regimes of these parks. While the former is jointly managed by Parks Australia and A<u>n</u>angu, who also hold title to the park, the latter is managed solely by the NSW NPWS and owned by the NSW Government. The varying approaches to interpretation within these two national parks are thus closely linked to land and power relations between Australian governments and Aboriginal peoples, highlighting the way that questions of control over land and Aboriginal sovereignty are entwined with how the cultural heritage of Aboriginal peoples is interpreted to the public.

However, existing joint management arrangements in Australian national parks are far from perfect, suggesting that sole Aboriginal management, if properly resourced, might be a preferable arrangement (Hill 2011; Senior Advisory Group 2021; tebrakunna country et al. 2016). This may be why an A<u>n</u>angu Traditional Custodian of Ulu<u>r</u>u-Kata Tju<u>t</u>a National Park suggested in an interview with me, that an Aboriginal controlled organisation such as the Central Land Council[1] might represent a viable alternative managing agency to Parks Australia. This organisation already manages a number of Indigenous Protected Areas[2] in the NT in consultation with their Traditional Custodians (Central Land Council n.d.). Although a similar management arrangement may be an option in Ulu<u>r</u>u-Kata Tju<u>t</u>a National Park at the expiry of the current 99-year lease with the Australian Government, my research suggests that this model, if applied more broadly, would not be supported by all Aboriginal peoples. For example, Traditional Custodians of the Blue Mountains expressed concerns that their status as Traditional Custodians was not recognised by the local Aboriginal Land Council responsible for this region. It follows from this that they would be unlikely to support the sole management of the Blue Mountains National Park by that agency. These differences suggest that a streamlined, uniform approach across all Australian protected areas would not suit the Aboriginal peoples I consulted, and that such decisions would need to be made on a case-by-case basis, by local Aboriginal peoples in consultation with government and other relevant agencies.[3]

As a direct result of consultations with Traditional Custodians of the study locales, my research reveals that decolonising heritage interpretation also requires rethinking Eurocentric assumptions that take public access to national parks as a given and prioritise tourism and recreation paradigms over Aboriginal people's sovereign relationships with Country. Although the Ulu<u>r</u>u climb has now closed, visitors continue to disregard cultural restrictions within the park by photographing culturally sensitive parts of Ulu<u>r</u>u where signage indicates that this is prohibited (Whittington and Waterton 2021). Similarly, Red Hands Cave was included within the Blue Mountains National Park and opened to the public without the consent of the Dharug and Gundungurra peoples. Red Hands Cave's "discovery" by settlers in 1913 resulted in the removal of cultural objects present in the cave by white anthropologists and local "collectors", and a long history of defacement of the cultural markings within the cave (Cameron 2014; Cawthorne 2016). In one Dharug man's view, revising the current interpretive signage at Red Hands Cave to include a Traditional Custodian voice and perspective would simply authorise the settler-colonial usurpation of this place by making it seem as if the Dharug people supported its inclusion within a national park and open public access to it.

My consultations with Traditional Custodians of the Blue Mountains indicate that questions of access to and interpretation of Red Hands Cave are a complex matter, on which there were a range of divergent perspectives. While some Dharug and Gundungurra peoples supported revised interpretive signage for this place, others expressed a preference for this place to be interpreted verbally by Traditional Custodians who accompanied visitors to the cave. Similarly, some Traditional Custodians favoured open access to the cave by the public, including visiting Aboriginal peoples, while others had concerns about non-Aboriginal people visiting the cave unaccompanied. These concerns were related to the disrespectful attitudes of some visitors and the history of deliberate damage to Red Hands Cave itself and to other places of cultural importance to Aboriginal peoples in the Sydney region. Recognising unceded Aboriginal sovereignty over Country means that decision-making about whether such places are open to the public and whether they are interpreted to them, and how, must be taken by Traditional Custodians and other Aboriginal peoples, and not left to the NSW NPWS to effectively designate such places as a "sacrificial" site.[4] The role of an agency such as the NSW NPWS in such a scenario would, I suggest, be as a mediator and facilitator, rather than a decision-maker.

My consultations with Blue Mountains' Traditional Custodians and my participation in cultural tours such as the Patji tour on APY lands also revealed that although visitors can experience aspects of Country, as non-Aboriginal people, the ontology of Country is not fully available to them. While visitors may come to appreciate some of the meanings of Country to A<u>n</u>angu, and to the Dharug and Gundungurra peoples, they do not have the ancestral ties, relationships with human and non-human kin, knowledge or responsibilities that would enable them to have an equivalent relationship with Country (Moreton-Robinson 2022; Mutu 2022). Forms of visitor Indigenisation such as placing ochred hand-prints on trees, as occurred at the Yellomundee Culture Camp, do not extinguish or replace Aboriginal sovereignty, nor make it available to non-Aboriginal peoples to extract. My research thus adds to the body of scholarship in cultural geography that casts doubt upon the legitimacy of practices of settler Indigenisation, including the appropriation of Indigenous ontologies and epistemologies by non-Indigenous peoples (Chandler and Reid 2019; Radcliffe 2017; Todd 2016). By addressing these issues, I highlight that Makarrata and reconciliation involve respect for what may well be incommensurable differences between Aboriginal and non-Aboriginal Australians, regardless of how much settlers such as myself may come to know Aboriginal places, including our own backyards, and, in my case, the Lomandra that grows there.

Visitors' desire for Indigenisation through cultural tourism is linked with "settler structures of feeling" (Rifkin 2011, 350), affective states capable of undermining the agency of Aboriginal peoples and stymying visitors' own "transformations". Anxiety in an Aboriginal cultural tourism context was one of the affective states reported by settler visitors, as opposed to international visitors, and was often associated with guilt about the negative impact of settler-colonialism on both the Australian landscape and Aboriginal peoples. Such guilt was often based on the belief that the colonisation of Australia had inflicted permanent and irreparable damage upon Aboriginal peoples and Country. These affective states inspired a deficit discourse defining Aboriginal identity in terms of dysfunction and lack, negating the plurality of Aboriginal identities and the resiliency of Aboriginal peoples.

A further prominent affective state among some visitors was anger, directed at climbers within Uluṟu-Kata Tjuṯa National Park and at the NSW NPWS for the problematic interpretive signage at Red Hands Cave. As these responses arose directly from experiences of the moment, such as witnessing visitors climbing Uluṟu, they could be said to emerge in interaction with other people and the environment, in place and time. Affect in cultural tourism is thus not simply a pleasurable experience of embodied connection with human and non-human others, but could also involve uncomfortable and challenging feelings, which nevertheless had the potential to change the visitor. Consequently, my research adds to the weight of scholarship within the fields of heritage and tourism studies that demonstrates that affect and emotion are not only important in constituting the visitor experience, but that feelings such as shame and empathy can propel visitors in the direction of critical, reflexive transformation (Witcomb 2015b, c). These changes, I argue, contra Smith (2014), often do more than simply confirm what visitors know or believe, altering them in ways they could not have predicted.

Although visitors' affective responses were not pre-determined by discourse, they were often influenced by it. As a consequence, affect, as I have argued throughout, cannot be regarded as separate from discourse. Affective states, such as anger, triggered by Uluṟu climbers, did not arise in a vacuum but had a meaning in the context of Australian settler-colonialism and the positionalities of visitors in relation to it. In illustrating these connections through a mixed methods approach, I engage with critical debates within the humanities and social sciences on the relative utility of more-than-representational theory and discourse theory (and aligned methodologies) in understanding visitors' experiences at heritage and tourism places. I do this by demonstrating that in visitor encounters, the more-than-representational is informed by the discursive rather than being separate from it, just as cognition is a function of emotion and embodiment. I argue, however, that visitor engagements with places of Aboriginal cultural heritage are far from predetermined, as they are also produced by the agentic human and non-human others with whom they share Country.

Additionally, my research contributes to the body of work in the social sciences on employing decolonising research methodologies. There were a number of ways I employed this approach. The research was conceptually orientated to the perspectives of Traditional Custodians as it arose from their public statements on their hopes for tourism on their traditional lands, and took as a primary focus the Aboriginal ontology of Country. In recognition of their unceded sovereignty, Traditional Custodians were consulted about whether the research should proceed, what form it should take, and the research locales where it should occur. Further orientating the research to the priorities of Traditional Custodians involved substantial changes to the initial research proposal, including targeting additional groups of research participants, including new research activities, altering the research locales chosen and revising planned timeframes. I also adopted specific research methodologies well-suited to working with Aboriginal peoples such as the methodology of yarning, and instituted additional steps to protect traditional knowledge by providing interview transcripts to Aboriginal interviewees for checking. The status of some Traditional Custodians as important knowledge holders was recognised by naming them in the research, at their request. Through a number of these processes, Traditional Custodians became decision-makers and stakeholders in the research, rather than simply participants, disrupting the unequal power dynamics between academic researcher and (Indigenous) research subject.

That Traditional Custodians had a stake in the research was apparent from requests to move my Uluṟu-based research forward so it could inform the development of a new Interpretation Strategy for Uluṟu-Kata Tjuṯa National Park, and the interest in my research informing interpretation planning for both Red Hands Cave and the Jenolan Caves. I shared de-identified research findings with research stakeholders to inform this future work, including producing a detailed report for Parks Australia and Aṉangu to assist with the Interpretation Strategy. I was thus able to demonstrate reciprocity with research stakeholders by sharing research findings and reorientating the research to potentially benefit Aboriginal communities, which is an important principle of decolonising research (Copenace et al. 2022; Smith 2021).

Although I was able to implement aspects of a decolonising methodology, unfortunately my work fell well short of the ideal, as my research did not fully engage Traditional Custodians of the research locales, nor was it fully controlled by them. Although I attempted to consult both Aṉangu and Gundungurra peoples on the research questions, a planned consultation with Aṉangu was cancelled due to Sorry Business. The time constraints imposed by a doctoral candidature and the limited funding available to support it, meant that I could not postpone the commencement of visitor interviews until Aṉangu were available to speak to me. I also attempted to consult Gundungurra Traditional Custodians on the research questions through the Indigenous Land Use Agreement (ILUA) Committee and to involve them in analysis of the research findings; however, they declined this level of participation. The departure of a member of this Committee who had acted as an informal advisor and advocate for the research (and was also a member of the Dharug community), made it difficult to resolve this issue. Needless to say, this situation was disappointing and I would have preferred a higher level of involvement from Traditional Custodians in all aspects of the research. However, in the case of the Gundungurra, it was important to respect their evident refusal to participate more fully in the research as their choice to make (Coulthard 2014). By being open about what worked well and the difficulties I faced, I hope to inform the decision-making of other researchers who also wish to engage with Indigenous and decolonising research methodologies.

Final thoughts

My consultations with Traditional Custodians of the research locales strongly suggest that cultural tourism does not only provide benefits to visitors but fulfils important functions for the Aboriginal peoples who provide it, quite separately to the economic benefits that can be the focus of government tourism policy (Ruhanen et al. 2015). Participation in cultural tourism enabled Anangu to fulfil their responsibilities as Traditional Custodians by teaching visitors how to behave on Country and aided the Traditional Custodians who offered the Yellomundee Culture Camp to resist colonisation through revived cultural practices. Aboriginal cultural tourism thus has the capacity to not only "transform" visitors but to fulfil the self-defined cultural, social and political goals of the Aboriginal peoples who offer it. This is not to deny that uncontrolled visitation to places of importance to Aboriginal peoples has caused injury and offense, and continues to do so. However, Aboriginal cultural tourism offers hope that it is possible to do "tourism otherwise to colonial logics" (Everingham et al. 2021, 2), and through it to assert continuing Aboriginal and Torres Strait Islander sovereignty over Country.

Notes

1 Aboriginal Land Councils are Aboriginal controlled organisations constituted in each Australian jurisdiction under various Land Rights Acts. Their key function, as established under the NSW Aboriginal Land Rights Act 1983, for example, is to acquire land, manage land that is acquired and to protect the culture and heritage of Aboriginal peoples on this land, https://legislation.nsw.gov.au/view/html/inforce/current/act-1983-042#sec.52 (accessed 22/11/2022).
2 An Indigenous Protected Area (IPA) is "an area of Indigenous-managed land or sea voluntarily dedicated by its traditional owners to be managed to promote biodiversity and cultural resource conservation" (Commonwealth of Australia 2012, 10).
3 Different legal arrangements for the management of protected areas across Australian jurisdictions would also problematise a streamlined approach, although I discuss this issue from the perspective of principle rather than practice.
4 The term "sacrificial site" was used by a NSW NPWS staff member to refer to Red Hands Cave in an interview with me, when I pointed out that even the trees and rocks leading from the car park to the Cave had been subject to graffiti and other forms of deliberate damage by visitors.

Bibliography

Abascal, T, Fluker, M & Jiang, M 2016, 'Domestic demand for Indigenous tourism in Australia: Understanding intention to participate', *Journal of Sustainable Tourism*, vol. 24, no. 8–9, pp. 1350–1368, DOI: 10.1080/09669582.2016.1193187.

Ablett, P & Dyer, P 2010, 'Heritage and hermeneutics: Towards a broader interpretation of interpretation', *Current Issues in Tourism*, vol. 12, no. 3, pp. 209–233.

Absolon, K & Willett, C 2005, 'Putting ourselves forward: Location in Aboriginal research', in L Brown and S Strega (eds), *Research as resistance: Critical, Indigenous and anti-oppressive approaches*, Canadian Scholars Press, Toronto, pp. 97–126, DOI: 10.1080/13683500802316063.

Ahmed, S 2004, *The cultural politics of emotion*, Routledge, New York, NY.

Ahmed, S 2005, 'The politics of bad feeling', *Australian Critical Race and Whiteness Studies Association Journal*, vol. 1, pp. 72–85, https://acrawsa.org.au/.

Ahmed, S 2007, 'A phenomenology of whiteness', *Feminist Theory*, vol. 8, no. 2, pp. 149–168, DOI: 10.1177/1464700107078139.

Ahmed, S 2010, 'Happy objects', in M Gregg & G Seigworth (eds), *The affect reader*, Duke University Press, Durham, NC, pp. 29–51.

Altman, J 2018, 'Raphael Lemkin in remote Australia: The logic of cultural genocide and homelands', *Oceania*, vol. 88, no. 3, pp. 336–359, DOI: 10.1002/ocea.5204.

Altman, J & Kerins, S 2012, *People on country – vital landscapes, Indigenous futures*, Federation Press, Annandale, NSW.

American National Park System Advisory Board 2006, *Charting a future for national heritage areas*, National Park System Advisory Board, Indianapolis.

Anderson, B & Harrison, P 2010, *Taking-place: Non-representational theories and geography*, Routledge, London.

ANTAR, n.d., *Australians for Native Title and Reconciliation*, https://antar.org.au/, accessed 11/10/2022.

Appleby, G & Davis, M 2018, 'The Uluru Statement and the promises of truth', *Australian Historical Studies*, vol. 49, no. 4, pp. 501–509, DOI: 10.1080/1031461X.2018.1523838.

Armitage, A 1995, *Comparing the policy of Aboriginal assimilation*, UBC Press, Australia, Canada, and New Zealand.

Ashenden, D 2010, 'Telling Tennant's story', *History Australia*, vol. 7, no. 3, pp. 52.1–52.15, DOI: 10.2104/ha100052.

Ashley, S & Stone, D 2023, 'Introduction on Stuart Hall and the imagining of heritage', in S Ashley & D Stone (eds), *Whose heritage?: Challenging race and identity in Stuart Hall's post-nation Britain*, ProQuest Ebook Central, pp. 1–10, http://ebookcentral.proquest.com/lib/uwsau/detail.action?docID=7245677.

Asten, M 2013, 'The Risdon Cove site: Birth of a state or site of a massacre; Bone of contention or future site of reconciliation?', *Tasmanian Historical Studies*, vol. 18, pp. 103–121.

Attwood, B 1992, 'Introduction', in B Attwood and J Arnold (eds), *Power, knowledge and Aborigines, La Trobe University Press in association with the National Centre for Australian Studies*, Monash University, Bundoora, pp. 1–16.

Australian Government 2020, *Seeing the Park through Anangu eyes: Interpretation Strategy 2020-2030*, viewed 13 October 2022, https://www.dcceew.gov.au.

Baird, M 2013, 'The breath of the mountain is my heart: Indigenous cultural landscapes and the politics of heritage', *International Journal of Heritage Studies*, vol. 19, no. 4, pp. 327–340, DOI: 10.1080/13527258.2012.663781.

Ballantyne, R 1995, 'Interpreters' conceptions of Australian Aboriginal culture and heritage: Implications for interpretive practice', *Journal of Environmental Education*, vol. 26, no. 4, pp. 11–17, DOI: 10.1080/00958964.1995.9941447.

Ballantyne, R, Packer, J & Bond, N 2012, 'Interpreting shared and contested histories: The broken links exhibition', *Curator- The Museum Journal*, vol. 55, no. 2, pp. 153–166, DOI: 10.1111/j.2151-6952.2012.00137.x.

Balvin, N & Kashima, Y 2012, 'Hidden obstacles to reconciliation in Australia: The persistence of stereotypes', in D Bretherton and N Balvin (eds), *Peace psychology in Australia*, Springer, New York, pp. 197–219, ProQuest Ebook Central, http://ebookcentral.proquest.com/lib/uwsau/detail.action?docID=884248.

Bandyopadhyay, R & Patil, V 2017, 'The white woman's burden' – The racialized, gendered politics of volunteer tourism', *Tourism Geographies*, vol. 19, no. 4, pp. 644–657, DOI: 10.1080/14616688.2017.1298150.

Banivanua Mar, T 2010, 'Craving wilderness: Queensland's national parks and the unsettling of emptied lands', in T Banivanua Mar & P Edmonds (eds), *Making settler colonial space - Perspectives on race, place and identity*, Palgrave Macmillan, Basingstoke, Hampshire, pp. 73–94.

Banner, S 2005, 'Why terra nullius? Anthropology and property law in early Australia', *Law and History Review*, vol. 23, no. 1, pp. 95–131, DOI: 10.1017/S0738248000000067.

Barnes, J 2011, 'Tourism's role in the struggle for the intellectual and material possession of 'The Centre' of Australia at Uluru, 1929-2011', *Journal of Tourism History*, vol. 3, no. 2, pp. 147–76, DOI: 10.1080/1755182X.2011.598575.

Bawaka Country, Suchet-Pearson, S, Wright, S, Lloyd, K & Burarrwang, L 2013, 'Caring as Country: Towards an ontology of co-becoming in natural resource management', *Asia Pacific Viewpoint*, vol. 54, no. 2, pp. 185–197, DOI: 10.1111/apv.12018.

Bawaka Country, Wright, S, Suchet-Pearson, S, Lloyd, K, Burarrwanga, L, Ganambarr, R, Ganambarr-Stubbs, M, Ganambarr, B & Maymuru, D 2015, 'Working with and learning from Country: decentring human author-ity', *Cultural Geographies*, vol. 22, no. 2, pp. 269–283, DOI: 10.1177/1474474014539248.

Bawaka Country, Wright, S, Lloyd, K, Suchet-Pearson, S, Burarrwanga, L, Ganambarr, R, Ganambarr, M & Ganambarr, B 2017, 'Meaningful tourist transformations with Country at Bawaka, North East Arnhem Land, northern Australia', *Tourist Studies*, vol. 17, no. 4, pp. 443–467, DOI: 10.1177/1468797616682134.

Beck, W & Somerville, M 2002, 'Embodied places in Indigenous ecotourism: The Yarrawarra research project', *Australian Aboriginal Studies*, vol. 2002, no. 2, pp. 4–13, https://search.informit.org/doi/10.3316/informit.593478174758093.

Belhassen, Y & Caton, K 2006, 'Authenticity matters', *Annals of Tourism Research*, vol. 33, no. 3, pp. 853–856, DOI: 10.1016/j.annals.2006.03.009.

Benjamin, W 1979, *Illuminations*, Translated by Harry Zohn and edited with an Introduction by Hannah Arendt, Fontana Press, London.

Bennett, C, Fitzpatrick-Harnish, K & Talbot, B 2022, 'Collaborative untangling of positionality, ownership, and answerability as white researchers in Indigenous spaces', *International Journal of Music Education*, vol. 40, no. 4, pp. 628–641, DOI: 10.1177/02557614221087343.

Berlant, L 2011, *Cruel optimism*, Duke University Press, ProQuest Ebook Central, http://ebookcentral.proquest.com/lib/uwsau/detail.action?docID=1172993.

Bessarab, D & Ng'andu, B 2010, 'Yarning about yarning as a legitimate method in Indigenous research', *International Journal of Critical Indigenous Studies*, vol. 3, no. 1, pp. 37–50, DOI: 10.5204/ijcis.v3i1.57.

Bhabha, H 2012, *The location of culture*, Taylor & Francis Group, ProQuest Ebook Central, http://ebookcentral.proquest.com/lib/uwsau/detail.action?docID=653022.

Bhambra, G 2014, 'Postcolonial and decolonial dialogues', *Postcolonial Studies*, vol. 17, no. 2, pp. 115–121, DOI: 10.1080/13688790.2014.966414.

Bird Rose, D 1996a, 'Land rights and deep colonising - The erasure of women', *Aboriginal Law Bulletin*, vol. 3, no. 85, pp. 6–13, http://classic.austlii.edu.au/au/journals/AboriginalLawB/1996/69.html.

Bird Rose, D 1996b, *Nourishing terrains- Australian Aboriginal views of landscape and wilderness*, Australian Heritage Commission, Canberra.

Bird Rose, D 2003, *Sharing kinship with nature: How reconciliation is transforming the NSW National Parks and Wildlife Service*, Australian National University, Canberra.

Bird Rose, D 2005, 'An Indigenous philosophical ecology- situating the human', *The Australian Journal of Anthropology*, vol. 16, no. 3, pp. 294–305, DOI: 10.1111/j.1835-9310.2005.tb00312.x.

Bird Rose, D 2011, *Country of the heart: An Indigenous Australian homeland*, Aboriginal Studies Press, Canberra Australia.

Bird Rose, D, James, D & Watson, C 2003, *Indigenous Kinship with the natural world in NSW*, NSW National Parks and Wildlife Service, Sydney.

Blaikie, N 2003, *Analyzing qualitative data*, Sage, London.

Blu Barnd, N 2017, *Native space- Geographic strategies to unsettle settler colonialism*, Oregon State University Press, Oregon.

Bondi, L, Davidson, J & Smith, M 2007, 'Introducing geography's emotional turn', in J Davidson, L Bondi and M Smith (eds), *Emotional Geographies*, Taylor and Francis, Abingdon, pp. 1–18.

Boudreau Morris, K 2017, 'Decolonizing solidarity: Cultivating relationships of discomfort', *Settler Colonial Studies*, vol. 7, no. 4, pp. 456–473, https://doi-org.ezproxy.uws.edu.au/10.1080/2201473X.2016.1241210.

Bowman, M 2015, *Every hill got a story: We grew up in country/men and women of Central Australia and the Central Land Council*, Hardie Grant Books, Richmond, Victoria.

Bowrey, K, Watson, I & Hadley, M 2022, 'Decolonising Aboriginal and Torres Strait Islander Research', *Australian Universities Review*, vol. 64, no. 1, pp. 54–64, https://search.informit.org/doi/10.3316/informit.485774526108850.

Bozoğlu, G 2020, *Museums, emotion and memory culture- The politics of the past in Turkey*, Routledge, Abingdon, Oxon and New York.

Braidotti, R 2019, 'A theoretical framework for the critical posthumanities', *Theory, Culture & Society*, vol. 36, no. 6, pp. 31–61, DOI: 10.1177/0263276418771486.

Breeden, S 1994, *Uluru: Looking after Uluru-Kata Tjuta the Anangu way*, Simon & Schuster, East Roseville, N.S.W.

Brigg, M & Graham, M 2020, 'The Relevance of Aboriginal political concepts: relationalism, not sovereignty', *ABC Religion and Ethics*, https://www.abc.net.au/religion/aboriginal-political-philosophy-relationalism/12954274, accessed April 14, 2021.

Brown, L & Strega, S 2005, 'Introduction- Transgressive possibilities', in L Brown & S Strega (eds), *Research as resistance: Critical, Indigenous and anti-oppressive approaches*, Canadian Scholars Press, Toronto, pp. 1–18.

Brown, S 2021, 'From culture and nature as separate to interconnected *naturecultures'*, in S Brown & C Goetcheus (eds), *The Routledge handbook of cultural landscape practice*, Taylor and Francis Limited, London, pp. 47–61.

Bruner, E 1991, 'Transformations of self in tourism', *Annals of Tourism Research*, vol. 18, no. 2, pp. 238–250, DOI: 10.1016/0160-7383(91)90007-X.

Bryant, M 2006, 'Tilden's children: Interpretation in Britain's national parks', in A Hems & M Blockley (eds), *Heritage interpretation*, Routledge, New York, pp. 173–189.

Buker, E 1990, 'Hidden desires and missing persons: A feminist deconstruction of Foucault', *A Western Political Quarterly*, vol. 43, no. 4, pp. 81–832.

Bunten, A 2018, 'Deriding demand: A case study of Indigenous imaginaries at an Australian Aboriginal tourism cultural park', in A Bunten & G Nelson (eds), *Indigenous tourism movements*, University of Toronto Press, Toronto, pp. 31–55.

Bunten, A & Graburn, N 2018, 'Current themes in Indigenous tourism', in A Bunten & G Nelson (eds), *Indigenous tourism movements*, University of Toronto Press, Toronto, pp. 1–27.

Byrne, D 1991, 'Western hegemony in archaeological heritage management', *History and Anthropology*, vol. 5, no. 2, pp. 269–276, DOI: 10.1080/02757206.1991.9960815.

Byrne, D 1996, 'Deep nation: Australia's acquisition of an Indigenous past', *Aboriginal History, vol.* 20, pp. 82–107, https://search.informit.org/doi/10.3316/ielapa.857526009992384.

Byrne, D 2003a, 'The Ethos of return: Erasure and reinstatement of Aboriginal visibility in the Australian historical landscape', *Historical Archaeology*, vol. 37, no. 1, pp. 73–86.

Byrne, D 2003b, 'Nervous landscapes: Race and space in Australia', *Journal of Social Archaeology*, vol. 3, no. 2, pp. 169–193, DOI: 10.1177/1469605303003002003.

Byrne, D 2014, *Counterheritage: Critical perspectives on heritage conservation in Asia*, Routledge, New York.

Byrne, D, Brockwell, S & O'Connor, S 2013, 'Introduction- engaging culture and nature', in S Brockwell, S O'Connor & D Byrne (eds), *Transcending the Culture–Nature divide in cultural heritage*, ANU Press, Canberra, pp. 1–13.

Cameron, B 2014, *A history of the Blue Labyrinth Blue Mountains National Park*, Bruce Cameron, Sun Valley.

Campbell, G, Smith, L & Wetherell, M 2017, 'Nostalgia and heritage: Potentials, mobilisations and effects', *International Journal of Heritage Studies*, vol. 23, no. 7, pp. 609–611, DOI: 10.1080/13527258.2017.1324558.

Carlson, B 2016, *The politics of identity: who counts as Aboriginal today?* AIATSIS, ProQuest Ebook Central, http://ebookcentral.proquest.com/lib/uwsau/detail.action?docID=4856127.

Carlson, B & Berglund, J 2021, *Indigenous peoples rise up: The global ascendancy of social media activism*, Rutgers University Press, New Brunswick.

Carpentier, N 2018, 'Discourse-theoretical analysis (DTA)', in *The Routledge Handbook of Critical Discourse Studies*, J. Flowerdew & J. E. Richardson (eds), Routledge, Oxon, pp, 272–284.

Carr, A 2004, 'Mountain places, cultural spaces: The interpretation of culturally significant Landscapes', *Journal of Sustainable Tourism*, vol. 12, no. 5, pp. 432–459, DOI: 10.1080/09669580408667248.

Carr, A, Ruhanen, L & Whitford, M 2016, 'Indigenous peoples and tourism: The challenges and opportunities for sustainable tourism', *Journal of Sustainable Tourism*, vol. 24, no. 8–9, pp. 1067–1079, DOI: 10.1080/09669582.2016.1206112.

Casey, ES 2001, 'Between geography and philosophy: What does it mean to be in the place-world?, *Annals of the Association of American Geographers*, vol. 91, no. 4, pp. 683–693, DOI: 10.1111/0004-5608.00266.

Cawthorne, M 2016, *Hidden history of the Blue Mountains*, Jimmy Dingo Publishing, Hazelbrook.

Central Land Council n.d, *Protected Areas*, viewed 23 November 2022, https://www.clc.org.au/protected-areas/.

Chambers, D & Buzinde, C 2015, 'Tourism and decolonisation: Locating research and self', *Annals of Tourism Research*, vol. 51, pp. 1–16, DOI: 10.1016/j.annals.2014.12.002.

Chandler, D & Reid, J 2019, *Becoming Indigenous – governing imaginaries in the Anthropocene*, Rowan and Littlefield, London.

Charles Walsh Nature Tourism Services 2001, *Greater Blue Mountains World Heritage Area Visitor Interpretation and Orientation Plan*, viewed 18 October 2022, http://www.naturetourismservices.com.au/images/detail-images/GBMWHA_pics/GBMWHA%20Interpretation%20Plan.pdf.

Charola, E & Meakins, F 2016, *Yijarni: True stories from Gurindji country*, Australian Institute of Aboriginal and Torres Strait Islander Studies, Canberra.

Chilisa, B 2012, *Indigenous research methodologies*, Sage Publications, California.

Clarke, A & Waterton, E 2015, 'A journey to the heart: Affecting engagement at Uluṟu-Kata Tjuṯa National Park', *Landscape Research*, vol. 40, no. 8, pp. 971–992, DOI: 10.1080/01426397.2014.989965.

Clifford, J 2013, *Returns: Becoming Indigenous in the Twenty-First Century*, Harvard University Press, Cambridge, Massachusetts.

Cole, D, Hammond, T & McCoo, S 1997, 'Information quantity and communication effectiveness: Low-impact messages on wilderness trail-side bulletin boards', *Leisure Sciences, vol.* 19, no. 1, pp. 59–72, DOI: 10.1080/01490409709512239.

Commonwealth of Australia 2012, *Respecting culture and country Indigenous Protected Areas in Australia: The first 15 years*, Canberra, Australian Government, viewed 23 November 2022, https://www.niaa.gov.au/sites/default/files/publications/respecting-culture-IPA.pdf.

Copenace, S, Cidro, J, Johnson, A & Anderson, K 2022, 'Auntie's Bundle-Conversation and research methodologies with Knowledge Gifter Sherry Copenace', in B Hokowhitu, A Moreton-Robinson, L Tuhiwai Smith, C Anderson & S Larkin (eds), *Routledge handbook of critical Indigenous studies*, Routledge, New York and London, pp. 189–202.

Coplan, A 2011, 'Understanding empathy- its features and effects', in A Coplan & P Goldie (eds), *Empathy: Philosophical and psychological perspectives*, Oxford University Press, Oxford, pp. 3–25.

Coplan, A & Goldie, P 2011, 'Introduction', in A Coplan and P Goldie (eds), *Empathy: Philosophical and psychological perspectives*, Oxford University Press, Oxford, pp. ix–xlvii.

Coulthard, G 2014, *Red skin, white masks: Rejecting the colonial politics of recognition*, University of Minnesota Press, Minneapolis.

Cowley, J & The Uluṟu Family 2018, *I am Uluṟu- A family's story*, Kungka Kutjara Aboriginal Corporation, Alice Springs.

Cowlishaw, G 1999, *Rednecks, eggheads and blackfellas: A study of racial power and intimacy in Australia*, Allen & Unwin, Sydney.

Cowlishaw, G 2009, *The city's outback*, UNSW Press, Sydney.

Cowlishaw, G 2010, 'Mythologising culture Part 1: Desiring Aboriginality in the suburbs', *The Australian Journal of Anthropology*, vol. 21, no. 2, pp. 208–227, DOI: 10.1111/j.1757-6547.2010.00079.x.

Cowlishaw, G 2011, 'Mythologising culture Part 2: Disturbing Aboriginality in the suburbs', *The Australian Journal of Anthropology*, vol. 22, no. 2, pp. 170–188, DOI: 10.1111/j.1757-6547.2011.00124.x.

Crang, M 2005, 'Travel/Tourism', in D Sibley, D Atkinson, P Jackson & N Washbourne (eds), *Cultural geography: A critical dictionary of key ideas, I.* B. Tauris & Company, Limited, https://ebookcentral.proquest.com/lib/uwsau/detail.action?docID=676401, pp. 34–40.

Crang, M & Tolia-Kelly, D 2010, 'Nation, race, and affect: Senses and sensibilities at national heritage sites', *Environment and Planning A*, vol. 42, no. 10, pp. 2315–2331, DOI: 10.1068/a4346.

Crouch, D 2010, 'Flirting with space: Thinking landscape relationally', *Cultural Geographies*, vol. 17, no. 1, pp. 5–18, DOI: 10.1177/1474474009349996.
Crouch, D 2015, 'Affect, heritage, feeling', in E Waterton & S Watson (eds), *Palgrave handbook of contemporary heritage research*, Palgrave Macmillan, Basingstoke, Hampshire, pp. 177–187.
Crowley, S 2022, 'Making visible the invisible: How combining autoethnography with visual arts practice unearthed more than I imagined', *Art/Research International: A Transdisciplinary Journal*, vol. 7, no. 1, pp. 156–185, DOI: 10.18432/ari29618.
Dalley, C 2020, 'The "White Card" is grey: Surveillance, endurance and the Cashless Debit Card', *Australian Journal of Social Issues*, vol. 55, no. 1, pp. 51–60, DOI: 10.1002/ajs4.100.
Daniel, Y. 1996, 'Tourism dance performances: Authenticity and creativity', *Annals of Tourism Research*, vol. 23, pp. 780–797.
Davison, G 1991, 'The meaning of heritage', in C Davison & C McConville (eds), *The heritage handbook*, Allen and Unwin, North Sydney, pp. 1–13.
Davidson, J & Milligan, C 2004, 'Embodying emotion, sensing space: Introducing emotional geographies', *Social and Cultural Geography*, vol. 5, no. 4, pp. 523–532, DOI: 10.1080/1464936042000317677.
de Bernardi, C 2019, 'A critical realist appraisal of authenticity in tourism: The case of the Sámi', *Journal of Critical Realism*, vol. 18, no. 4, pp. 437–452, DOI: 10.1080/14767430.2019.1654702.
de Simine, S 2013, *Mediating memory in the museum: Trauma, empathy, nostalgia*, Palgrave Macmillan, New York.
Deepak, C 2021, *Resilience, authenticity and digital heritage tourism*, Routledge, London.
Deleuze, G & Guattari, F 1987, *A thousand plateaus: Capitalism and schizophrenia*, Continuum, London.
Dewsbury, J, Harrison, P, Rose, M & Wylie, J 2002, 'Enacting geographies', *Geoforum*, vol. 33, no. 4, pp. 437–440, DOI: 10.1016/S0016-7185(02)00029-5.
d'Hauteserre, A 2015, 'Affect theory and the attractivity of destinations', *Annals of Tourism Research*, vol. 55, no. c, pp. 77–89, DOI: 10.1016/j.annals.2015.09.001.
Dirlik, A 1997, *The postcolonial aura*, Westview Press, Boulder.
Dittmer, J & Waterton, E 2019, '"You'll go home with bruises": Affect, embodiment and heritage on board HMS Belfast', *Area, Royal Geographic Society*, vol. 51, no. 4, pp. 706–718, DOI: 10.1111/area.12513.
Dodson, M 2003, 'The end in the beginning: Re(de)finding Aboriginality', in M Grossman (ed.), *Blacklines: Contemporary critical writings by Indigenous Australians*, Melbourne University Publishing, ProQuest Ebook Central, http://ebookcentral.proquest.com/lib/uwsau/detail.action?docID=5628344, pp. 32–37.
Duelke, B 2005, 'Knowing tradition, dealing with history? On concepts, strategies and practices', in L Taylor, G Ward, G Henderson, R Davis & L Wallis (eds), *The power of knowledge- the resonance of tradition*, Aboriginal Studies Press, Canberra, pp. 199–213.
Dudgeon, P & Bray, A 2019, 'Indigenous relationality: Women, kinship and the law', *Genealogy, vol.* 3, no. 2, pp. 23–34, DOI: 10.3390/genealogy3020023.
Duff, C 2010, 'On the role of affect and practice in the production of place', *Environment and Planning D: Society and Space*, vol. 28, no. 5, pp. 881–895, DOI: 10.1068/d16209.
Dunn, K, Forrest, J, Burnley, I & McDonald, A 2004, 'Constructing racism in Australia', *Australian Journal of Social Issues*, vol. 39, no. 4, pp. 409–430, DOI: 10.1002/j.1839-4655.2004.tb01191.x.
Dussel, E. 2008, 'Anti-Cartesian meditations: About the origin of the philosophical anti-discourse of modernity', *Tabula Rasa*, vol. 9, pp. 153–198.

Dussel, E 2012, 'Transmodernity and interculturality: An interpretation from the perspective of philosophy of liberation', *Transmodernity: Journal of Peripheral Cultural Production of the Luso-Hispanic World*, vol. 1, no. 3, pp, pp. 28–59, DOI: 10.5070/T413012881.

Edwards, B 2011, 'A personal journey with Anangu history and politics', *Flinders Journal of History and Politics*, vol. 27, pp. 13–25, https://www-proquest-com.ezproxy.uws.edu.au/docview/922557084?pq-origsite=primo&accountid=36155.

Erickson, B 2018, 'Anachronistic others and embedded dangers: Race and the logic of whiteness in nature tourism', in B Grimwood, K Canton & L Cooke (eds), *New moral natures in tourism*, Taylor & Francis Group, ProQuest Ebook Central, http://ebookcentral.proquest.com/lib/uwsau/detail.action?docID=5352033, pp. 43–56.

Evans, J, tebrakunna country & Lee, E 2022, 'Indigenous women's voices: 20 years on from Linda Tuhiwai Smith's Decolonizing methodologies', in tebrakunna country, E Lee and J Evans (eds), *Indigenous women's voices: 20 years on from Linda Tuhiwai Smith's Decolonizing methodologies*, Bloomsbury Publishing, London, pp18–32.

Everingham, P, Peters, A & Higgins-Desbiolles, F 2021, 'The (im)possibilities of doing tourism otherwise: The case of settler colonial Australia and the closure of the climb at Uluru', *Annals of Tourism Research*, vol. 88, pp. 1–11, DOI: 10.1016/j.annals.2021.103178.

Fairclough, N 2001, 'Critical discourse analysis as a method in social scientific research', in R Wodak and M Meyer (eds), *Methods of critical discourse analysis*, Sage Publications, London, pp 121–138.

Fairclough, N, 2003, *Analysing discourse: Textual analysis for social research*, Routledge, London.

Feshbach, N & Feshbach, S 2009, 'Empathy and education', in J Decety & W Ickes (eds) *The social neuroscience of empathy*, MIT Press, Cambridge, MA, pp. 85–97.

Fine, M, Weis, L, Powell, L C & Wong, L M (Eds.) 1997, *Off white: Readings on race, power, and society*. Taylor & Frances/Routledge, New York.

Finegan, C 2019, 'The interpreter as researcher: Ethical heritage interpretation in Indigenous contexts', *Journal of Heritage Tourism*, vol. 14, no. 3, pp. 282–294, DOI: 10.1080/1743873X.2018.1474883.

Fforde, C, Bamblett, L, Lovett, R, Gorringe, S & Fogarty, B 2013, 'Discourse, deficit and identity: Aboriginality, the race paradigm and the language of representation in contemporary Australia', *Media International Australia*, vol. 149, no. 1, pp. 162–173, DOI: 10.1177/1329878X1314900117.

Fogarty, W, Lovell, M, Langenberg, J & Heron, M 2018, Deficit discourse and strengths based approaches: Changing the narrative of Aboriginal and Torres Strait Islander health and well-being, The Lowitja Institute, Melbourne, viewed 2 December 2022, https://www.lowitja.org.au/page/services/resources/Cultural-and-social-determinants/racism/deficit-discourse-strengths-based.

Foley, G & Anderson, T 2006, 'Land rights and Aboriginal voices', *Australian Journal of Human Rights*, vol. 12, no. 1, pp. 83–108, DOI: 10.1080/1323238X.2006.11910814.

Foucault, M 1980, 'The confession of the flesh', in (eds), *Power/Knowledge - selected interviews and other writings 1972-1977*, Pantheon Books, New York, pp. 194–229.

Foucault, M 1981, 'The order of discourse', in R Young (ed), *Untying the text: A post-structuralist reader*, Routledge and Kegan Paul, London, pp. 48–78.

Foucault, M 1984a, 'The body of the condemned', in P Rainbow (ed), *The Foucault reader*, Pantheon Books, New York, pp. 170–178.

Foucault, M 1984b, 'The repressive hypothesis', in *The Foucault reader*, P Rainbow (ed), Pantheon Books, New York, pp. 301–330.

Foucault, M 2002, *The archaeology of knowledge*, (A. Sheridan Smith, Trans.), Routledge, London and New York.

Foxlee, J 2007, 'Cultural landscape interpretation: The case of the sorry rock story at Uluṟu-Kata Tjuṯa National Park', *Tourism Recreation Research*, vol. 32, no. 3, pp. 49–56, DOI: 10.1080/02508281.2007.11081539.

Frankenberg, R 1993, *White women, race matters: The social construction of whiteness* (NED-New edition), University of Minnesota Press, Minneapolis, http://www.jstor.org/stable/10.5749/j.ctttsnhh.

Fraser, J 2007, 'Museums- drama, ritual and power', in S Knell, S Macleod & S Watson (eds), *Museum revolutions: How museums change and are changed*, Routledge, Oxon, pp. 291–302.

Fredericks, B 2013, 'We don't leave our identities at the city limits': Aboriginal and Torres Strait Islander people living in urban localities', *Australian Aboriginal Studies (Canberra)*, vol. 2013, no. 1, pp. 4–16, https://search.informit.org/doi/10.3316/informit.422915104662241.

Galliford, M 2010, 'Touring 'Country', sharing 'home': Aboriginal tourism, Australian tourists and the possibilities for cultural transversality', *Tourist Studies*, vol. 10, no. 3, pp. 227–244, DOI: 10.1177/1468797611407759.

Galliford, M 2012, 'Travels in 'becoming-Aboriginal': Research reciprocating between anecdotes and theory', *Critical Arts*, vol. 26, no. 3, pp. 401–423, DOI: 10.1080/02560046.2012.705463.

Gapps, S 2018, *The Sydney wars*, NewSouth Publishing, Sydney.

Gapps, S 2021, *Gudyarra - The first Wiradyuri war of resistance*, NewSouth Publishing, Sydney.

Gelder, K & Jacobs, J 1998, *Uncanny Australia: Sacredness and identity in a postcolonial nation*, Melbourne University Press, Carlton, Vic.

Giraudo, R 2018, 'San cultural tourism: Mobilizing Indigenous agency in Botswana', in A Bunten & G Nelson (eds), *Indigenous tourism movements*, University of Toronto Press, Toronto, pp. 117–139.

Gokcigdem, E 2016, 'Introduction', in E Gokcigdem (ed.), *Fostering empathy through museums*, Rowman & Littlefield, Lanham, Maryland, pp. xix–xxxii.

Gonzalez, P & Vazquez, A 2014, 'Between planning and heritage: Cultural parks and national heritage areas', *European Spatial Research and Policy*, vol. 21, no. 2, pp. 33–46.

Gorringe, S, Ross, J & Fforde, C 2011, *Will the real Aborigine please stand up': Strategies for breaking the stereotypes and changing the conversation*, AIATSIS Research Discussion Paper No.28, Canberra, Australian Institute of Aboriginal and Torres Strait Islander Studies, viewed 2 December 2022, www.aiatsis.gov.au/research/discussion.html.

Gosse, W 1873, W C Gosse's Explorations, 1873, viewed 29 February 2024, https://gutenberg.net.au/ebooks13/1306451h.html.

Graham, B, Ashworth, G & Tunbridge, J 2000, *A geography of heritage: Power, culture and economy*, Arnold, London.

Graham, M 1999, 'Some thoughts about the philosophical underpinnings of Aboriginal worldviews', *Global Religions, Culture, and Ecology*, vol. 3, no. 2, pp. 105–118, DOI: 10.1163/156853599X00090.

Gregg, M & Seigworth, G 2010, *The affect theory reader*, Duke University press, Durham, NC.

Gregory, K & Witcomb, A 2007, 'Beyond nostalgia: The role of affect in generating historical understanding at heritage sites', in S Knell, S McLeod & S Watson (eds), *Museum Revolutions*, Routledge, New York, pp. 263–274.

Grieves, V 2008, 'The "battlefields": Identity, authenticity and Aboriginal knowledges in Australia', in E Minde (ed.), *Indigenous peoples: Self-determination, knowledge, identity*, Eboron Academic Publishers, The Netherlands, pp. 287–311.

Grimwood, B, Stinson, M & King, L 2019, 'A decolonising settler story', *Annals of Tourism Research*, vol. 79, loc 102763, DOI: 10.1016/j.annals.2019.102763.
Grosfoguel, R 2015, 'Epistemic Racism/Sexism, Westernized Universities and the Four Genocides/Epistemicides of the Long Sixteenth Century', in M Araújo & S Maeso (eds), *Eurocentrism, racism and knowledge*, Palgrave Macmillan, Basingstoke, UK, pp. 23–46.
Guattari, F 1995, *Chaosmosis*, Power Publication, Sydney, Australia.
Haebich, A 2011, 'Forgetting Indigenous histories: Cases from the history of Australia's Stolen Generations', *Journal of Social History*, vol. 44, no. 4, pp. 1033–1046, DOI: 10.1353/jsh.2011.0042.
Hage, G 2000, *White nation - Fantasies of white supremacy in a multicultural society*, Pluto Press, Annandale.
Haila, Y 2000, 'Beyond the nature-culture dualism', *Biology and Philosophy*, vol. 15, no. 2, pp. 155–175, DOI: 10.1023/A:1006625830102.
Hall, M 1994, *Tourism and politics: Policy, power and place*, Wiley, Chichester, UK.
Hall, M & Tucker, H 2004, *Tourism and postcolonialism: Contested discourses, identities and representations*, Routledge, London.
Hall, S 1992, 'The west and the rest', in *Formations of Modernity*, S Hall & B Gieben (eds), Polity Press, Cambridge, pp. 275–332.
Hall, S 1997, 'The spectacle of the 'Other', in S Hall (ed), *Cultural Representations and Signifying Practices*, The Open University, London, pp. 223–291.
Hall, S 1999, 'Un-settling 'the heritage', re-imagining the post-nation - Whose heritage?, *Third Text*, vol. 13, no. 49, pp. 3–13, DOI: 10.1080/09528829908576818.
Hall, S 2001, 'Foucault: Power, knowledge and discourse', in M Wetherell, S Taylor & S Yates (eds), *Discourse theory and practice: A reader*, Sage Publications, London, pp. 72–81.
Hall, S 2021, 'Race, the floating signifier: What more is there to say about "race" [1997]', in P Gilroy & R Wilson Gilmore (eds), *Stuart hall- selected writings on race and difference*, Duke University Press, Durham.
Hall, S, Hobson, D, Lowe, A & Willis, P 2005, *Culture, media, language - Working papers in cultural studies, 1972–79*, Routledge, London.
Harrison, R 2004, *Shared landscapes - Archaeologies of attachment and the pastoral industry in NSW*, UNSW Press, Sydney.
Harrison, R 2013, *Heritage: Critical approaches*, Routledge, Milton Park, Abingdon; New York.
Harrison, R 2014, 'Shared histories: Rethinking 'Colonized' and 'Colonizer' in the archaeology of colonialism', in N Ferris, R Harrison & M Wilcox (eds), *Rethinking colonial pasts through archaeology*, Oxford University Press, pp. 37–56.
Harrison, R 2015, 'Beyond "Natural" and "Cultural" heritage: Toward an ontological politics of heritage in the age of Anthropocene', *Heritage & Society*, vol. 8, no. 1, pp. 24–42, DOI: 10.1179/2159032X15Z.00000000036.
Harvey, D 2001, 'Heritage pasts and heritage presents: Temporality, meaning and the scope of heritage studies', *International Journal of Heritage Studies*, vol. 7, no. 4, pp. 319–338, DOI: 10.1080/13581650120105534.
Harvey, D 2015, 'Landscape and heritage: Trajectories and consequences', *Landscape Research*, vol. 40, no. 8, pp. 911–924, DOI: 10.1080/01426397.2014.967668.
Healy, S & Mulcahy, D 2021, 'Pedagogic affect: Assembling an affirming ethics', *Pedagogy, Culture & Society*, vol. 29, no. 4, pp. 555–572, DOI: 10.1080/14681366.2020.1768581.
Hemmings, C 2005, 'Invoking affect: Cultural theory and the ontological turn', *Cultural Studies*, vol. 19, no. 5, pp. 548–567, DOI: 10.1080/09502380500365473.
Higgins-Desbiolles, F 2003, 'Reconciliation tourism: Tourism healing divided societies!' *Tourism Recreation Research*, vol. 28, no. 3, pp. 35–44, DOI: 10.1080/02508281.2003.11081415.

Higgins-Desbiolles, F, Doering, A & Bigby, B 2022, *Socialising tourism - Rethinking tourism for social and ecological justice*, Routledge, London and New York.
Hill, L 2015, 'Human geography and archaeology: Strange bedfellows?, *Progress in Human Geography*, vol. 39, no. 4, pp. 412–431, DOI: 10.1177/0309132514521482.
Hill, R 2011, 'Towards equity in Indigenous co-management of protected areas: Cultural planning by Miriuwung-Gajerrong people in the Kimberley, Western Australia', *Geographical Research*, vol. 49, no. 1, pp. 72–85, DOI: 10.1111/j.1745-5871.2010.00669.x.
Hinkson, M 2018, 'Turbulent dislocations in Central Australia: Exile, place making, and the promises of elsewhere', *American Ethnologist*, vol. 45, no. 4, pp. 521–532, DOI: 10.1111/amet.12706.
Hinkson, M & Fullenwieder, L 2019, 'Imaging crisis in Indigenous Australia and Canada: Towards an analysis of neoliberal primitivism', *Visual Studies*, vol. 34, no. 2, pp. 164–181, DOI: 10.1080/1472586X.2019.1640636.
Hoare, J 2020, 'The practice and potential of heritage emotion research: An experimental mixed-methods approach to investigating affect and emotion in a historic house', *International Journal of Heritage Studies*, vol. 26, no. 10, pp. 955–974, DOI: 10.1080/13527258.2020.1714696.
Hodge, B & Mishra, V 1990, *Dark side of the dream: Australian literature and the postcolonial mind*, Allen and Unwin, Sydney.
Hokowhitu, B, Moreton-Robinson, A, Tuhiwai Smith, L, Anderson, C & Larkin, S 2022, *Routledge handbook of critical Indigenous studies*, Routledge, New York and London.
Horáková, M 2022, 'From landscape to country: Writing settler belonging in post-Mabo Australia', *Life Writing*, vol. 19, no. 2, pp. 295–314, DOI: 10.1080/14484528.2020.1781584.
Howitt, R, Doohan, K, Suchet-Pearson, S, Cross, S, Lawrence, R, Lunkapis, G, Muller, S, Prout, S & Veland, S 2013, 'Intercultural capacity deficits: Contested geographies of coexistence in natural resource management', *Asia Pacific Viewpoint*, vol. 54, no. 2, pp. 126–140, DOI: 10.1111/apv.12014.
Hueneke, H & Baker, R 2009, 'Tourist behaviour, local values and interpretation at Uluṟu: The sacred deed at Australia's mighty heart', *GeoJournal*, vol. 74, no. 5, pp. 477–90, DOI: 10.1007/s10708-008-9249-2.
Huggins, J, Huggins, R & Jacobs, J 1995, 'Kooramindanjie: Place and the postcolonial', *History Workshop Journal*, vol. 39, pp. 164–181.
Hughes, M & Morrison-Saunders, A 2002, 'Impact of trail-side interpretive signs on visitor knowledge', *Journal of Ecotourism*, vol. 1, no. 2–3, pp. 122–132, DOI: 10.1080/14724040208668119.
Hughes, M & Morrison-Saunders, A 2005, 'Influence of on-site interpretation intensity on visitors to natural areas', *Journal of Ecotourism*, vol. 4, no. 3, pp. 161–177, DOI: 10.1080/jJET.v4.i3.pg161.
Hvenegaard, G, Marshall, H & Lemelin, R 2016, 'Hot interpretation of controversial topics at Batoche National Historic Site, Saskatchewan, Canada', *Journal of Interpretation Research*, vol. 21, no. 2, pp. 45–62, DOI: 10.1177/109258721602100204.
Ingold, T 1993, 'The temporality of the landscape', *World Archaeology*, vol. 25, no. 2, pp. 152–17, DOI: 10.1080/00438243.1993.9980235.
Jäger, S & Maier, F 2016, 'Analysing discourses and dispositives: A Foucauldian approach to theory and methodology', in R Wodak & M Meyer (eds), *Methods of critical discourse studies*, 3rd Edition, Sage, London, pp. 109–137.
James, D 2015, 'Tjukurpa time', in A McGrath & M Jebb (eds), *Long history, deep time- Deepening histories of place*, ANU Press, Canberra, pp. 33–45.
James, S 2007, 'Constructing the climb: Visitor decision-making at Uluṟu', *Geographical Research*, vol. 45, no. 4, pp. 398–407, DOI: 10.1111/j.1745-5871.2007.00478.x.

Johnston, E 1991, *Royal Commission into Aboriginal deaths in custody - National Report*, vol. *1*, viewed 15 June 2022, http://www.austlii.edu.au/au/other/IndigLRes/rciadic/national/vol1/.

Jones, A & Jenkins, K 2014, 'Rethinking collaboration: Working the Indigene-colonizer hyphen', in N Denzin, Y Lincoln Y & L Tuhiwai Smith (eds), *Handbook of critical and Indigenous methodologies*, Sage Publications, London, pp. 471–486.

Jørgensen, M & Phillips, L. 2011, *Discourse analysis as theory and method*, Sage Publications, DOI: 10.4135/9781849208871.

Kantor, I 2007, 'Ethnic cleansing and America's creation of national parks', *Public Land and Resources Law Review*, vol. 28, pp. 42–62.

Karskens, G 2009, *The Colony: A history of early Sydney*, Allen and Unwin, Sydney.

Karskens, G 2020, *People of the river*, Allen and Unwin, Sydney.

Kearney, A 2009, 'Homeland emotion: An emotional geography of heritage and homeland', *International Journal of Heritage Studies*, vol. 15, no. 2–3, pp. 209–222, DOI: 10.1080/13527250902890746.

Kidd, J & Sayner, B 2019, 'Intersections of silence and empathy in heritage practice', *International Journal of Heritage Studies*, vol. 25, no. 1, pp. 1–4, DOI: 10.1080/13527258.2018.1475414.

Kizuk, S 2020, 'Settler shame: A critique of the role of shame in Settler–Indigenous relationships in Canada', *Hypatia*, vol. 35, no. 1, pp. 161–77, DOI: 10.1017/hyp.2019.8.

Klein, E 2020, 'Settler colonialism in Australia and the cashless debit card', *Social Policy Administration*, vol. 54, no. 2, pp. 265–277, DOI: 10.1111/spol.12576.

Knudsen, B & Waade, A 2010, 'Performative authenticity in tourism and spatial experience: Rethinking the relations between travel, place and emotion', in B Knudsen & M Waade (eds), *Re-investing authenticity: Tourism, place and emotions*, Channel View Publications, pp. 1–20, ProQuest Ebook Central, http://ebookcentral.proquest.com/lib/uwsau/detail.action?docID=543894.

Knudsen, B & Stage, C 2015, *Affective methodologies: Developing cultural research strategies for the study of affect*, Palgrave Macmillan, London.

Knudsen, B & Andersen, C 2019, 'Affective politics and colonial heritage, Rhodes Must Fall at UCT and Oxford', *International Journal of Heritage Studies*, vol. 25, no. 3, pp. 239–258, DOI: 10.1080/13527258.2018.1481134.

Knudsen, B, Oldfield, J, Buettner, E & Zabunyan, E 2022, 'Introduction', in B Knudsen, J Oldfield, E Buettner & E Zabunyan (eds), *Decolonizing colonial heritage*, Routledge, London and New York, pp. 1–21.

Kolig, E 2005, 'The politics of Indigenous-or ingenious-tradition: Some thoughts on the Australian and New Zealand situation', in T Otto & P Pedersen (eds), *Tradition and agency: Tracing cultural continuity and invention*, Aarhus University Press, Aarhus, pp. 293–326.

Kølvraa, C & Knudsen, B 2020, 'Decolonizing European colonial heritage in urban spaces – An introduction to the special issue', *Heritage & Society*, vol. 13, no. 1–2, pp. 1–9, DOI: 10.1080/2159032X.2021.1888370.

Koopman, W, Watling, C & LaDonna, K 2020, 'Autoethnography as a Strategy for Engaging in Reflexivity', *Global Qualitative Nursing Research*, vol. 7, pp. 1–9, https://doi.org/10.1177/2333393620970508.

Kovach, M 2005, 'Emerging from the margins: Indigenous methodologies', in L Brown & S Strega (eds), *Research as resistance: Critical, Indigenous and anti-oppressive approaches*, Canadian Scholars Press, Toronto, pp. 19–36.

Kluttz, J, Walker, J & Walter, P 2020, 'Unsettling allyship, unlearning and learning towards decolonising solidarity', *Studies in the Education of Adults*, vol. 52, no. 1, pp. 49–66, DOI: 10.1080/02660830.2019.1654591.

Laclau, E & Mouffe, C 1985, *Hegemony and socialist strategy: Towards a radical democratic politics*, Verso, London.
Lähdesmäki, T 2017, 'Politics of affect in the EU Heritage policy discourse: An analysis of promotional videos of sites awarded with the European Heritage Label', *International Journal of Heritage Studies*, vol. 23, no. 8, pp. 709–722, DOI: 10.1080/13527258.2017.1317649.
Lambert-Pennington, K 2012, '"Real Blackfellas": Constructions and meanings of urban Indigenous identity', *Transforming Anthropology*, vol. 20, no. 2, pp. 131–145, DOI: 10.1111/j.1548-7466.2012.01154.x.
Langton, M 1996, 'What do we mean by wilderness? Wilderness and Terra Nullius in Australian art', *The Sydney Papers*, vol. 8, no. 1, pp. 11–31, https://search.informit.org/doi/10.3316/agispt.19961755.
Langton, M 2018, *Welcome to Country - A travel guide to Indigenous Australia*, Hardie Grant Travel, Sydney.
Lattas, A 1992, 'Primitivism, nationalism and individualism in Australian popular culture', *Journal of Australian Studies*, vol. 16, no. 35, pp. 45–58, https://doi.org/10.1080/14443059209387117.
Lean, G 2016, *Transformative travel in a mobile world*, CAB International, viewed 28 September 2020, ProQuest Ebook Central, http://ebookcentral.proquest.com/lib/uwsau/detail.action?docID=4412680.
Lean, G, Staiff, R & Waterton, E 2014, 'Exploring travel and transformation' in G Lean & R Staiff (eds), *Travel and transformation*, viewed 25 September 2020, ProQuest Ebook Central, https://ebookcentral.proquest.com/lib/uwsau/detail.action?docID=1711849, pp. 11–25.
Leckie, E 2021, 'Federal government to roll out $30 million job-ready package for cashless debit card sites', *ABC News*, viewed 10 June 2021, https://www.abc.net.au/news/2021-05-04/cashless-debit-card-trial-sites-catered-job-package-services/100113824.
Lee, E 2016, 'Protected Areas, country and value: The nature–culture tyranny of the IUCN's Protected Area Guidelines for Indigenous Australians', *Antipode*, vol. 48, no. 2, pp. 355–374, DOI: 10.1111/anti.12180.
Lentin, A 2020, *Why race still matters*, Polity Press, Cambridge.
Liebelt, B 2019, 'Touching grindstones in archaeological and cultural heritage practice: Materiality, affect and emotion in settler-colonial Australia', *Australian Archaeology*, vol. 85, no. 3, pp. 267–278, DOI: 10.1080/03122417.2019.1751982.
Lincoln, Y, Lynham, S & Guba, E 2018, 'Paradigmatic controversies, contradictions, and emerging confluences, revisited', in N Denzin and Y Lincoln (eds), *The Sage handbook of qualitative research*, 5th edition, Sage Publications, Thousand Oaks, pp. 212–263.
Linehan, D, Clark, I & Xie, P 2020, *Colonialism, tourism and place - Global transformations in tourist destinations*, Edward Elgar Publishing, Cheltenham, UK.
Ling Tan, P, Noor, S, Rasoolimanesh, M & Mustafa, H 2020, 'Communication and visitor factors contributing towards heritage visitors' mindfulness', *Journal of Heritage Tourism*, vol. 15, no. 1, pp. 27–43, DOI: 10.1080/1743873X.2019.1586909.
Little, A 2020, 'The politics of Makarrata: Understanding Indigenous–Settler relations in Australia', *Political Theory*, vol. 48, no. 1, pp. 30–56, DOI: 10.1177/0090591719849023.
Little, A & McMillan, M 2017, 'Invisibility and the politics of reconciliation in Australia: Keeping conflict in view', *Ethnopolitics*, vol. 16, no. 5, pp. 519–537, DOI: 10.1080/17449057.2016.1219473.
Littler, J 2005, 'Introduction- British heritage and the legacies of 'race'', in J Littler and R Naidoo (eds), *The politics of heritage: The legacies of race*, Taylor & Francis Group, accessed 3 October 2023, ProQuest Ebook Central, http://ebookcentral.proquest.com/lib/uwsau/detail.action?docID=199418, pp. 1–17.

Lloyd, K, Suchet-Pearson, S, Wright, S, Tofa, M, Rowland, C, Burarrwanga, L, Ganambarr, R, Ganambarr, M, Ganambarr, B & Maymuru, D 2015, 'Transforming tourists and "culturalising commerce": Indigenous tourism at Bawaka in Northern Australia', *The International Indigenous Policy Journal*, vol. 6, no. 4, pp. 1–21, DOI: 10.18584/iipj.2015.6.4.6.

Lorimer, H 2005, 'Cultural geography: The busyness of being 'more-than representational'', *Progress in Human Geography*, vol. 29, no. 1, pp. 83–94, DOI: 10.1191/0309132505ph531pr.

Lovell, J & Hitchmough, S 2020, *Authenticity in North America: Place, tourism, heritage, culture and the popular imagination*, Routledge/Taylor & Francis Group, London, New York, NY.

Lowenthal, D 1998, *The heritage crusade and the spoils of history*, Cambridge University Press, New York.

Lugones, M 2007, 'Heterosexualism and the colonial/modern gender system', *Hypatia*, vol. 22, no. 1, pp. 186–209, DOI: 10.1111/j.1527-2001.2007.tb01156.x.

McGrath, A 2015, 'Crossing history's mountains: The historic values of the Greater Blue Mountains', in D Benson (ed) *Values for a New Generation: Greater Blue Mountains World Heritage Area*, Greater Blue Mountains World Heritage Area Advisory Committee, Katoomba, pp. 96–135.

Macdonald, G 2001, 'Does 'culture' have 'history'? Thinking about continuity and change in central New South Wales', *Aboriginal History*, vol. 25, pp. 176–199, DOI: 10.22459/AH.25.2011.12.

Mackey, E 1998, 'Becoming Indigenous: Land, belonging, and the appropriation of Aboriginality in Canadian nationalist narratives', *Social Analysis: The International Journal of Anthropology*, vol. 42, no. 2, pp. 150–178, https://www.jstor.org/stable/23166573.

Mackey, E 2014, 'Unsettling expectations: (Un)certainty, settler states of feeling, law, and decolonization', *Canadian Journal of Law and Society*, vol. 29, no. 2, pp. 235–252, DOI: 10.1017/cls.2014.10.

Mackey, E 2016, *Unsettled expectations: Uncertainty, land and settler decolonization*, Fernwood Publishing, Halifax and Winnipeg.

Macpherson, H 2010, 'Non-representational approaches to body-landscape relations', *Geography Compass*, vol. 4, no. 1, pp. 1–13, DOI: 10.1111/j.1749-8198.2009.00276.x.

Maddison, S 2012, 'Postcolonial guilt and national identity: Historical injustice and the Australian settler state', *Social Identities: Journal for the Study of Race, Nation and Culture*, vol. 18, no. 6, pp. 695–709, DOI: 10.1080/13504630.2012.709000.

Maddison, S 2013, 'Indigenous identity, 'authenticity' and the structural violence of settler colonialism', *Identities*, vol. 20, no. 3, pp. 288–303, DOI: 10.1080/1070289X.2013.806267.

Maddison, S 2022, *'How do we tell the truth about Australia', On-line presentation sponsored by Australia Centre*, Melbourne University, delivered on 16 March 2022.

Madgin, R & Lesh, J 2021, *People-centred methodologies for heritage conservation: Exploring emotional attachments to historic urban places*, Routledge, Abingdon, Oxon, New York, NY.

Maeso S and Araújo M 2015, 'Eurocentrism, political struggles and the entrenched will-to-ignorance: an introduction', in Araújo and S Maeso (eds), *Eurocentrism, racism and knowledge*, Palgrave Macmillan, Basingstoke, UK, pp. 1–22.

Mahadevan, R 2018, 'Examining domestic and international visits in Australia's Aboriginal tourism', *Tourism Economics*, vol. 24, no. 1, pp. 127–134, DOI: 10.1177/1354816617701440.

Maldonado-Torres, N 2007, 'On the coloniality of being', *Cultural Studies*, vol. 21, no. 2–3, pp. 240–270, DOI: 10.1080/09502380601162548.

Marchant, A 2019, *Historicising heritage and emotions: The affective histories of blood, stone and land*, Routledge, London.
Martin, K 2010, 'Living pasts- contested tourism authenticities', *Annals of Tourism Research*, vol. 37, no. 2, pp. 537–554, DOI: 10.1016/j.annals.2009.11.005.
Martin, K & Mirraboopa, B 2003, 'Ways of knowing, being and doing: A theoretical framework and methods for Indigenous and Indigenist re-search', *Journal of Australian Studies*, vol. 27, no. 76, pp. 203–214, DOI: 10.1080/14443050309387838.
Mason, R & Sayner, J 2019, 'Bringing museal silence into focus: Eight ways of thinking about silence in museums', *International Journal of Heritage Studies*, vol. 25, no. 1, pp. 5–20, DOI: 10.1080/13527258.2017.1413678.
Massumi, B 2002, *Parables for the virtual: Movement, affect, sensation*, Duke University Press, Durham, NC.
McGloin, C 2015, 'Listening to hear: Critical allies in Indigenous Studies', *Australian Journal of Adult Learning*, vol. 55, no. 2, pp. 267–282, https://search.informit.org/doi/10.3316/aeipt.209717.
McKittrick, K 2021, *Dear science and other stories*, Duke University Press, Durham.
McRainey, D & Russick, J 2010, *Connecting kids to history with museum exhibitions*, Left Coast Press, Walnut Creek, CA.
Metro-Roland, M & Soica, S 2019, 'Landscape and tourism', in E Waterton and M Attha (eds) *The Routledge companion to landscape studies*, Routledge, Abingdon, Oxon; New York, NY, pp. 354–365.
Mezirow, J 1978, 'Perspective transformation', *Adult Education*, vol. 28, no. 2, pp. 100–110, DOI: 10.1177/074171367802800202.
Mezirow, J 2000, *Learning as transformation: Critical perspectives on a theory in progress*, Jossey-Bass, San Francisco, CA.
Micieli-Voutsinas, J 2017, 'An absent presence: Affective heritage at the National September 11th Memorial & Museum', *Emotion, Space and Society*, vol. 24, pp. 93–104, DOI: 10.1016/j.emospa.2016.09.005.
Mignolo, W 2007, 'Delinking', *Cultural Studies*, vol. 21, no. 2–3, pp. 449–514, DOI: 10.1080/09502380601162647.
Mignolo, W 2011, 'Geopolitics of sensing and knowing: On (de)coloniality, border thinking and epistemic disobedience', *Postcolonial Studies*, vol. 14, no. 3, pp. 273–283, DOI: 10.1080/13688790.2011.613105.
Mills, S 1997, *Discourse*, Routledge, London.
Mills, S 2003, *Michel Foucault*, Routledge, London.
Mkono, M 2012, 'Authenticity does matter', *Annals of Tourism Research*, vol. 39, no. 1, pp. 480–483, DOI: 10.1016/j.annals.2011.06.004.
Mkono, M 2013, 'African and western tourists- object authenticity quest?', *Annals of Tourism Research*, vol. 41, pp. 195–214, http://dx.doi.org/10.1016/j.annals.2013.01.002.
Moran, A 2002, 'As Australia decolonizes: Indigenizing settler nationalism and the challenges of settler/Indigenous relations', *Ethnic and Racial Studies*, vol. 25, no. 6, pp. 1013–1042, DOI: 10.1080/0141987022000009412.
Moran, A 2005, 'White Australia, settler nationalism and Aboriginal assimilation', *Australian Journal of Politics and History*, vol. 51, no. 2, pp. 168–193, DOI: 10.1111/j.1467-8497.2005.00369.x.
Moran, A 2009, 'What settler Australians talk about when they talk about Aborigines: Reflections on an in-depth interview study', *Ethnic and Racial Studies*, vol. 32, no. 5, pp. 781–801, DOI: 10.1080/01419870802023936.
Moreton-Robinson, A. 1999, 'Unmasking Whiteness: A Goori Jondal's Look at Some Duggai Business', in B McKay (ed.), *Unmasking Whiteness: Race relations and reconciliation, Queensland studies*, Centre for Public Culture and Ideas, Griffith University, Brisbane, pp. 28–36, https://doi.org/10.1017/s1321816600001823.

Moreton-Robinson, A 2003, 'Tiddas talkin' up to the white woman: When Huggins et. al. took on Bell', in M Grossman (ed.), *Blacklines: Contemporary critical writing by Indigenous Australians*, Melbourne University Press, Melbourne, Victoria, pp. 66–77.

Moreton-Robinson, A 2015, *The white possessive: Property, power and Indigenous sovereignty*, University of Minnesota Press, Minneapolis.

Moreton-Robinson, A 2016, 'Race and cultural entrapment: Critical Indigenous studies in the twenty-first century, in A Moreton-Robinson (ed.) *Critical Indigenous studies: Engagements in first world locations*, University of Arizona Press, Tuscon, pp. 102–115.

Moreton-Robinson, A 2022, 'Incommensurable sovereignties- Indigenous ontology matters', in B Hokowhitu, A Moreton-Robinson, L Tuhiwai Smith, C Anderson & S Larkin (eds), *Routledge handbook of critical Indigenous studies*, Routledge, New York and London, pp. 257–268.

Muir, S 2014, 'Real people: Authenticity and Aboriginality in the Australian holistic milieu', *Ethnos*, vol. 79, no. 4, pp. 473–495, DOI: 10.1080/00141844.2013.770411.

Muller, S, Hemming, S & Rigney, D 2019, 'Indigenous sovereignties: Relational ontologies and environmental management', *Geographical Research*, vol. 57, no. 4, pp. 399–410, DOI: 10.1111/1745-5871.12362.

Munro, J, Morrison-Saunders, A & Hughes, M 2008, 'Environmental interpretation evaluation in natural Areas', *Journal of Ecotourism*, vol. 7, no. 1, pp. 1–14, DOI: 10.2167/joe137.0.

Mutu, M 2022, 'Mana Maori Motuhake- Maori concepts and practices of sovereignty', in B Hokowhitu, A Moreton-Robinson, L Tuhiwai Smith, C Anderson & S Larkin (eds), *Routledge handbook of critical Indigenous studies*, Routledge, New York and London, pp. 269–282.

National Health and Medical Research Council 2018, *Ethical conduct in research with Aboriginal and Torres Strait Islander Peoples and communities: Guidelines for researchers and stakeholders*, Commonwealth of Australia, Canberra.

Nayar, P 2010, *Postcolonialism - A guide for the perplexed*, Continuum International Publishing Group, London.

Ngai, S 2005, *Ugly feelings*, Harvard University Press, Cambridge, Massachusetts.

Nelson, J 2020, 'Intimate interventions: Responses to racist talk in families', *Emotion, Space and Society*, Vol. 35, pp. 1–7, DOI: 10.1016/j.emospa.2020.100676.

Nicholls, R 2009, 'Research and Indigenous participation: Critical reflexive methods', *International Journal of Social Research Methodology*, vol. 12, no. 2, pp. 117–126, DOI: 10.1080/13645570902727698.

Nilsen, A & Bader, M 2016, 'The psychology of empathy: Compelling possibilities for museums', in E Gokcigdem (ed.), *Fostering empathy through museums*, Rowman & Littlefield, Lanham, Maryland, pp. 115–130.

Ngurra, D, Dadd, L, Glass, P, Scott, R, Graham, M, Judge, S, Hodge, P & Suchet-Pearson, S 2019, 'Yanama budyari gumada: Reframing the urban to care as Darug Country in Western Sydney', *Australian Geographer*, vol. 50, no. 3, pp. 279–293, DOI: 10.1080/00049182.2019.1601150.

NSW Government n.d.a, *Crown Land*, viewed 9 August 2022, https://reservemanager.crownland.nsw.gov.au.

NSW Government n.d.b, *State Heritage Inventory Jenolan Caves*, viewed 25 November 2022, https://www.hms.heritage.nsw.gov.au/App/Item/ViewItem?itemId=5051578.

NSW Government 2001, *Blue Mountains National Park Plan of Management 2001*, viewed 18 October 2022, https://www.environment.nsw.gov.au/research-and-publications/publications-search/blue-mountains-national-park-plan-of-management.

NSW Government 2009a, *Yellomundee Regional Park Plan of Management*, viewed 18 October 2022, https://www.environment.nsw.gov.au/research-and-publications/publications-search/yellomundee-regional-park-plan-of-management.

NSW Government 2009b, *Greater Blue Mountains World Heritage Area Strategic Plan*, viewed 19 October 2022, https://www.environment.nsw.gov.au/research-and-publications/publications-search/greater-blue-mountains-world-heritage-area-strategic-plan.

NSW Government 2017, *Jenolan Karst Conservation Reserve Conservation Management Plan 2017*, document not publicly available.

NSW Government 2019, *Jenolan Karst Conservation Reserve Plan of Management 2019*, viewed 18 October 2022, https://www.environment.nsw.gov.au/research-and-publications/publications-search/jenolan-karst-conservation-reserve-plan-of-management.

NSW National Parks and Wildlife Service n.d., *Red Hands Cave*, viewed 17 April 2020, https://www.nationalparks.nsw.gov.au/things-to-do/aboriginal-sites/red-hands-cave.

Palomino-Schalscha, M 2018, 'Beyond neoliberalism and nature: Territoriality, relational ontologies, and hybridity in a tourism initiative in Alto Bio-Bio, Chile', in A Bunten & G Nelson (eds), *Indigenous tourism movements*, University of Toronto Press, Toronto, pp. 222–241.

Palmer, C 1994, 'Tourism and colonialism - The experience of The Bahamas', *Annals of Tourism Research*, vol. 21, no. 4, pp. 792–811, DOI: 10.1016/0160-7383(94)90084-1.

Paradies, Y 2006, 'Beyond black and white-essentialism, hybridity and Indigeneity', *Journal of Sociology*, vol. 42, no. 4, pp. 355–367, DOI: 10.1177/1440783306069993.

Parks Australia n. d. a, *Data on tourist visitation to Uluṟu*, personal email correspondence from Parks Australia.

Parks Australia n. d. b, *Fact Sheet – Tjukurpa*, https://uluru.gov.au/static/58c59795b85f20c2a46409965fb6416c/uktnp-document-fs-tjukurpa.pdf, accessed, 25 November 2024.

Parry, B 2004, *Postcolonial studies - A materialist critique*, Routledge, London and New York.

Pearson, L 2017, 'What is a Makarrata? The Yolngu word is more than a synonym for treaty', *ABC News Online*, viewed 12 October 2022, https://www.abc.net.au/news/2017-08-10/makarrata-explainer-yolngu-word-more-than-synonym-for-treaty/8790452.

Picca, L & Feagin, J 2007, *Two-faced racism: Whites in the backstage and frontstage*, Routledge, New York.

Pile, S 2010, 'Emotions and affect in recent human geography', *Transactions of the Institute of British Geographers*, vol. 35, no. 1, pp. 5–20, DOI: 10.1111/j.1475-5661.2009.00368.x.

Pink, S 2015, *Doing sensory ethnography*, Sage Publications, London.

Plater, S 2017, 'Bulldust, flat tyres and roadkill: A disorderly decolonizing fieldwork journey through remote Aboriginal and Torres Strait Islander Australia', *Australian Aboriginal Studies*, vol 2, pp. 70–83, https://search.informit.org/doi/10.3316/agispt.20180043.

Pomering, A & White, E 2011, 'The portrayal of Indigenous identity in Australian tourism brand advertising: Engendering an image of extraordinary reality or staged authenticity?' *Place Branding and Public Diplomacy*, vol. 7, pp. 165–174, DOI: 10.1057/pb.2011.19.

Porter, L 2010, *Unlearning the colonial cultures of planning*, Ashgate publishing, Surrey, England.

Porter, L 2018, 'From an urban Country to urban Country: Confronting the cult of denial in Australian cities', *Australian Geographer*, vol. 49, no. 2, pp. 239–246, DOI: 10.1080/00049182.2018.1456301.

Pratt, M 1992, *Imperial eyes: Travel writing and transculturation*, Routledge, London.

Probyn, E 2005, *Blush-faces of shame*, University of Minnesota Press, accessed 3 February 2021, ProQuest Ebook Central http://ebookcentral.proquest.com/lib/uwsau/detail.action?docID=310698.

Prout, S 2011, 'Urban myths: Exploring the unsettling nature of Aboriginal presence in and through a regional Australian town', *Urban Policy and Research*, vol. 29, no. 3, pp. 275–291, DOI: 10.1080/08111146.2011.578300.

Prout, S & Howitt, R 2009, 'Frontier imaginings and subversive Indigenous spatialities', *Journal of Rural Studies*, vol. 25, no. 4, pp. 396–403, DOI: 10.1016/j.jrurstud.2009.05.006.

Punch, S & Tisdall, E 2012, 'Exploring children and young people's relationships across Majority and Minority Worlds', *Children's Geographies*, vol. 10, no. 3, pp. 241–248, DOI: 10.1080/14733285.2012.693375.

Purvis, T & Hunt, A 1993, 'Discourse, ideology, discourse, ideology, discourse, ideology...', *The British Journal of Sociology*, vol. 44, no. 3, pp. 473–499, DOI: 10.2307/591813.

Quayson, A 2000, *Postcolonialism, theory, practice or process?* Polity Press, Cambridge.

Queensland University of Technology 2021, *Recommended Guidelines for Aboriginal and Torres Strait Islander Terminology*, viewed 28 November 2022, https://cms.qut.edu.au/__data/assets/pdf_file/0008/477143/qut-guidelines-for-aboriginal-torres-strait-islander-terminology.pdf.

Quijano, A 2007, 'Coloniality and modernity/rationality', *Cultural Studies*, vol. 21, no. 2–3, pp. 168–178, DOI: 10.1080/09502380601164353.

Quin, R 2017, *Orientalism - An analysis of Edward said's orientalism*, Macat International Limited, Milton.

Radcliffe, S 2017, 'Geography and Indigeneity I: Indigeneity, coloniality and knowledge', *Progress in Human Geography*, vol. 41, no. 2, pp. 220–229, DOI: 10.1177/0309132515612952.

Rickly-Boyd, J 2013, 'Existential authenticity: Place matters', *Tourism Geographies*, vol. 15, no. 4, pp. 680–686, DOI: 10.1080/14616688.2012.762691.

Rifkin, M 2011, 'Settler states of feeling: National belonging and the erasure of native American presence', in C Levander & R Levine (eds), *Companion to American literary studies*, John Wiley & Sons, Incorporated, ProQuest Ebook Central, http://ebookcentral.proquest.com/lib/uwsau/detail.action?docID=4041671, pp. 350–363.

Rifkin, M 2013, 'Settler common sense', *Settler Colonial Studies*, vol. 3, no. 3–4, pp. 322–340, DOI: 10.1080/2201473X.2013.810702.

Rix, E, Barclay, L & Wilson, S 2014, 'Can a white nurse get it? 'Reflexive practice' and the non-Indigenous clinician/researcher working with Aboriginal people', *Rural and Remote Health*, vol. 14, no. 2, pp. 2679–2679, DOI: 10.22605/RRH2679.

Robin, L 2014, 'Wilderness in a global age, fifty years on', *Environmental History*, vol. 19, no. 4, pp. 721–727, DOI: 10.1093/envhis/emu072.

Roccas, S, Klan, Y & Liviatan, I 2004, 'Exonerating cognitions, group identification and personal predictors of collective guilt among Jewish-Israelis', in N Branscombe & B Doosje (eds), *Collective guilt: International perspectives*, Cambridge University Press, Cambridge, pp. 130–147.

Rose, G 1997, 'Situating knowledges: Positionality, reflexivities and other tactics', *Progress in Human Geography*, vol. 21, no. 3, pp. 305–320, DOI: 10.1191/030913297673302122.

Rounds, J 2006, 'Doing identity work in museums', *Curator*, vol. 49, no. 2, pp. 133–150, DOI: 10.1111/j.2151-6952.2006.tb00208.x.

Ruhanen, L, Whitford, M & McLennan, C 2015, 'Indigenous tourism in Australia: Time for a reality check', *Tourism Management*, vol. 48, pp. 73–83, DOI: 10.1016/j.tourman.2014.10.017.

Ruhanen, L & Whitford, M 2019, 'Cultural heritage and Indigenous tourism', *Journal of Heritage Tourism*, vol. 14, no. 3, pp. 179–191, DOI: 10.1080/1743873X.2019.1581788.

Russell-Mundine, G 2012, 'Reflexivity in Indigenous research: Reframing and decolonising research?, *Journal of Hospitality and Tourism Management*, vol. 19, no. 1, pp. 85–90, DOI: 10.1017/jht.2012.8.

Ryan, L, Pascoe, W, Debenham, J, Gilbert, S, Richards, J, Smith, R, Owen, C, Anders, R, Brown, M, Price, D, Newley, J & Usher, K. (n. d.), *Colonial Frontier Massacres Australia 1780-1930*, viewed 18 May 2020, https://c21ch.newcastle.edu.au/colonialmassacres/map.php.

Said, E 1978, *Orientalism*, Vintage Books, New York.

Salazar, N & Graburn, H 2014, 'Introduction - Toward an anthropology of tourism imaginaries' in N Salazar & H Graburn, *Tourism imaginaries, Berghahn Books*, New York, NY, pp. 1–30.

Sather-Wagstaff, J 2017, 'Making polysense of the world: Affect, memory, heritage', in D Tolia-Kelly, E Waterton & S Watson (eds), *Heritage, affect and emotion: Politics, practices and infrastructures*, Routledge, New York, pp. 12–29.

Saukko, P 2011, *Doing research in cultural studies*, Sage Publications, London.

Saul, H & Waterton, E 2018, 'Rethinking frontiers: Transformation, exploration and adventure', in H Saul and E Waterton (eds), *Affective geographies of transformation, exploration and adventure*, Routledge, London, pp. 1–14.

Savenije, G & de Bruijn, P 2017, 'Historical empathy in a museum: Uniting contextualisation and emotional engagement', *International Journal of Heritage Studies*, vol. 23, no. 9, pp. 832–845, DOI: 10.1080/13527258.2017.1339108.

Schorch, P, Watson, S & Waterton, E 2017, 'Museum canopies and affective cosmopolitanism: Cultivating cross-cultural landscapes for ethical, embodied responses' in D Tolia-Kelly, E Waterton & S Watson (eds), *Heritage, affect and emotion: Politics, practices and infrastructures*, Routledge, New York, pp. 93–113.

Schwandt, T 2015, *The Sage dictionary of qualitative inquiry*, Sage, Thousand Oaks.

Scollon, R 2001, 'Action and text: Towards an integrated understanding of the place of text in social (inter)action, mediated discourse analysis and the problem of social action', in R Wodak and M Meyer (eds), *Methods of critical discourse analysis*, Sage Publications, London, pp 139–184.

Searle, T & Muller, S 2019, 'Whiteness" and natural resource management: Let's talk about race baby, let's talk about sovereignty!', *Geographical Research*, vol. 57, no. 4, pp. 411–424, DOI: 10.1111/1745-5871.12379.

Seiver, B & Matthews, A 2016, 'Beyond whiteness: A comparative analysis of representations of Aboriginality in tourism destination images in New South Wales, Australia', *Journal of Sustainable Tourism*, vol. 24, no. 8–9, pp. 1298–1314, DOI: 10.1080/09669582.2016.1182537.

Senior Advisory Group 2021, Joint Management Arrangements for Commonwealth National Parks, viewed 13 October 2022, https://www.dcceew.gov.au/parks-heritage/national-parks/parks-australia/publications/sag-joint-management-arrangements-commonwealth-national-parks-advice.

Shome, R 1999, 'Whiteness and the politics of location,' in T. Nakayama & G. Martin (eds), *Whiteness: The communication of social identity*, Sage Publications, California, pp. 107–128.

Silverman, H 2015, 'Heritage and authenticity' in Waterton, E & Watson, S (eds), *The Palgrave handbook of contemporary heritage research*, Palgrave Macmillan, UK, pp. 69–88, ProQuest Ebook Central, http://ebookcentral.proquest.com/lib/uwsau/detail.action?docID=1953015.

Simpson, M 2006, 'Revealing and concealing: Museums, objects, and the transmission of knowledge in Aboriginal Australia", in J Marstine (ed.), *New museum theory and practice: An introduction*, Blackwell Publishing, Oxford, pp. 153–174.

Sissons, J 2005, *First peoples: Indigenous cultures and their futures*, Reaktion Books, London.

Slack, S 2021, *Interpreting heritage: A guide to planning and practice*, Routledge, Abingdon, Oxon.

Slater, L 2019, *Anxieties of belonging in settler colonialism*, Routledge, New York and London.

Slote, M 2007, *The ethics of care and empathy*, Routledge, London and New York, https://doi-org.ezproxy.uws.edu.au/10.4324/9780203945735.

Smith, A, McNiven, I, Rose, D, Brown, S, Johnston, C & Crocker, S 2019, 'Indigenous knowledge and resource management as world heritage values: Budj Bim cultural landscape, Australia', *Archaeologies*, vol. 15, pp. 285–313, https://doi-org.ezproxy.uws.edu.au/10.1007/s11759-019-09368-5.

Smith, C 2021, 'Pastoral place-making: Rock art as a social agent in colonial encounters', *Rock Art Research*, vol. 38, no. 2, pp. 195–210.

Smith, L 2004, *Archaeological theory and the politics of cultural heritage* (1st ed.), Routledge. DOI: 10.4324/9780203307991.

Smith, L 2006, *The uses of heritage*, Routledge, Abingdon, Oxon.

Smith, L 2012, 'A pilgrimage of masculinity: The Stockman's Hall of Fame and Outback Heritage Centre', *Australian Historical Studies*, vol. 43, no. 3, pp. 472–482, DOI: 10.1080/1031461X.2012.708116.

Smith, L 2014, 'Visitor emotion, affect and registers of engagement at museums and heritage Sites', *Conservation Science in Cultural Heritage,* vol. 14, no. 2, pp. 125–131, DOI: 10.6092/issn.1973-9494/5447.

Smith, L 2021, *Emotional heritage: Visitor engagement at museum and heritage sites*, Routledge, Abingdon, Oxon.

Smith, L & Campbell, G 2015, 'The elephant in the room: Heritage, affect, and emotion', in M Logan, M Nic Craith & U Kockel (eds), *A companion to heritage studies*, John Wiley and Sons, New Jersey, pp. 443–460.

Smith, LT 2021, *Decolonizing methodologies: Research and Indigenous peoples*, 3rd Edition, Bloomsbury Publishing, New York.

Smith, V 1978, *Hosts and guests: The anthropology of tourism*, Basil Blackwell, Oxford.

Spivak, G 1988, 'Can the subaltern speak', in C Nelson & L Grossburg (eds), *Marxism and the interpretation of culture*, Macmillan Education, London, pp. 271–313.

Spivak, G 1990, 'In a word; interview', *Differences*, vol. 1, no. 2, pp. 124–155.

Staiff, R 2014, *Re-imagining heritage interpretation: Enchanting the past-future*, Taylor and Francis, London.

Staiff, R, Bushell, R & Kennedy, P 2002, 'Interpretation in national parks: Some critical questions', *Journal of Sustainable Tourism*, vol. 10, no. 2, pp. 97–113, DOI: 10.1080/09669580208667156.

Stanner, W 2009, *The dreaming and other essays*, Black Inc. Agenda, Melbourne.

Steiner, C & Reisinger, Y 2006, 'Understanding existential authenticity', *Annals of Tourism Research*, vol. 33, no. 2, pp. 299–318, DOI: 10.1016/j.annals.2005.08.002.

Stinson, M & Grimwood, B 2019, 'On actor-network theory and anxiety in tourism research', *Annals of Tourism Research*, vol. 77, pp. 141–143,

Stinson, M, Grimwood, B & Caton, K 2021, 'Becoming common plantain: Metaphor, settler responsibility, and decolonizing tourism', *Journal of Sustainable Tourism*, vol. 29, no. 2–3, pp. 234–252, DOI: 10.1080/09669582.2020.1734605.

Stocker, K 2018, 'The alchemy of tourism: From stereotype and marginalizing discourse to real in the space of tourist performance', in A Bunten & G Nelson (eds), *Indigenous tourism movements*, University of Toronto Press, Toronto, pp. 73–95.

Stronza, A 2001, 'Anthropology of tourism: Forging new ground for ecotourism and other alternatives', *Annual Review of Anthropology*, vol. 30, no. 1, pp. 261–283, DOI: 10.1146/annurev.anthro.30.1.261.

tebrakunna country, Lee, E & Tran, T 2016, 'From boardroom to kitchen table: Shifting the power seat of Indigenous governance in protected area management', *Australian Aboriginal Studies*, vol. 2, pp. 81–93, https://search.informit.org/doi/10.3316/agispt.20170804.

tebrakunna country & Lee, E 2022, 'Reclaiming the first person voice', in tebrakunna country, E Lee and J Evans (eds), *Indigenous women's voices: 20 years on from*

Linda Tuhiwai Smith's Decolonizing Methodologies, Bloomsbury Publishing, London, pp. 149–161.

Theodossopoulos, D 2013a, 'Laying claim to authenticity: Five anthropological dilemmas', *Anthropological Quarterly*, vol. 86, no. 2, pp. 337–360, DOI: 10.1353/anq.2013.0032.

Theodossopoulos, D 2013b, 'Emberá Indigenous tourism and the trap of authenticity: Beyond inauthenticity and invention', *Anthropological Quarterly*, vol. 86, no. 2, pp. 397–426, DOI: 10.1353/anq.2013.0023.

Theodossopoulos, D 2018, 'Indigenous tourism as a transformative process: The case of the Embara in Panama', in A Bunten & G Nelson (eds), *Indigenous tourism movements*, University of Toronto Press, Toronto, pp. 99–116.

Thien, D 2005, 'After or beyond feeling? A consideration of affect and emotion in geography', *Area*, vol. 37, no. 4, pp. 450–454, DOI: 10.1111/j.1475-4762.2005.00643a.x.

Thrift, N. 1996, *Spatial formations*, London, Sage.

Thrift, N 1997, 'The still point', in S Pile and M Keith (eds), *Geographies of resistance*, Routledge, London, pp. 124–151.

Thrift, N 2004, 'Intensities of feeling: Towards a spatial politics of affect', *Geografiska Annaler: Series B, Human Geography*, vol. 86, no. 1, pp. 57–78, DOI: 10.1111/j.0435-3684.2004.00154.x.

Thrift, N 2008, *Non-representation theory: Space, politics, affect*, Routledge, London.

Tilden, F 2007, *Interpreting our heritage, Fourth Edition*, The University of North Carolina Press, North Carolina.

Todd, Z 2016, 'An Indigenous feminist's take on the ontological turn: 'ontology' is just another word for colonialism', *Journal of Historical Sociology*, vol. 29, no. 1, pp. 4–22, DOI: 10.1111/johs.12124.

Tolia-Kelly, D 2006, 'Affect – An ethnocentric encounter? Exploring the 'universalist' imperative of emotional/affectual geographies', *Area*, vol. 38, no. 2, pp. 213–217, DOI: 10.1111/j.1475-4762.2006.00682.x.

Tolia-Kelly, D 2016, 'Feeling and being at the (postcolonial) museum', *Sociology*, vol. 50, no. 5, pp. 896–912, DOI: 10.1177/0038038516649554.

Tolia-Kelly, D, Waterton, E & Watson, S 2017, *Heritage, affect and emotion: Politics, practices and infrastructures*, Routledge, New York.

Tomkins, S 2008, *Affect, imagery, consciousness*: The complete edition *Volumes 1 and 11*, Springer, New York.

Trask, H-K 1991, 'Natives and anthropologists: The colonial Struggle', *The Contemporary Pacific*, vol. 3, no. 1, pp. 159–167.

Trask, H-K 1999, *From a native daughter: Colonialism and sovereignty in Hawai'i*, University of Hawai'i Press, Honolulu.

Travesi, C 2018, 'Knowing and being known. Approaching Australian Indigenous tourism through Aboriginal and non-Aboriginal politics of knowing', *Anthropological Forum*, vol. 28, no. 3, pp. 275–292, DOI: 10.1080/00664677.2018.1486285.

Trigger, D, Mulcock, J, Gaynor, A & Toussaint, Y 2008, 'Ecological restoration, cultural preferences and the negotiation of 'nativeness' in Australia', in', *Geoforum*, vol. 39, no. 3, pp. 1273–1283, DOI: 10.1016/j.geoforum.2007.05.010.

Tuck, E & Yang, K 2012, 'Decolonization is not a metaphor', Decolonization: *Indigeneity, Education & Society*, vol. 1, no. 1, pp. 1–40, https://jps.library.utoronto.ca/index.php/des/article/view/18630/15554.

Tuck, E, Gaztambide-Fernandez, R & Ruben, A 2013, 'Curriculum, replacement, and settler futurity', *Journal of Curriculum Theorizing*, vol. 29, no. 1, pp. 72–89, https://www.proquest.com/docview/1436933689/fulltextPDF/A1CB1C847C9B4123PQ/1?accountid=36155.

Tuck, E & McKenzie, M 2015, *Place in research: Theory, methodology, and methods*, Routledge, New York.

Tucker, H & Shelton, E 2018, 'Tourism, mood and affect: Narratives of loss and hope', *Annals of Tourism Research*, vol. 70, pp. 66–75, DOI: 10.1016/j.annals.2018.03.001.

Tynan, L 2021, 'What is relationality? Indigenous knowledges, practices and responsibilities with Kin', *Cultural Geographies*, vol. 28, no. 4, pp. 597–610, DOI: 10.1177/14744740211029287.

United Nations Committee on Social, Economic and Cultural Rights 2009, *General Comment No. 21 Right of Everyone to Take Part in Cultural Life*, viewed 9 October 2016, http://hrlibrary.umn.edu/gencomm/escgencom21.html.

UNESCO 2019, *Operational Guidelines for the Implementation of the World Heritage Convention*, UNESCO World Heritage Centre, viewed 12 October 2022, https://whc.unesco.org/en/guidelines/.

Urry, J 2011, *The tourist gaze 3.0: Leisure and travel in contemporary societies*, Third Edition, Sage, London.

Uzzell, D 1989, *Heritage interpretation*, Belhaven Press, London.

Van Winkle, C & Backman, K 2011, 'Designing interpretive audio tours to enhance meaningful learning transfer at a historic site', *Journal of Heritage Tourism*, vol. 6, no. 1, pp. 29–43, DOI: 10.1080/1743873X.2010.518761.

Veracini, L 2015, *The settler-colonial present*, Palgrave Macmillan, London.

Veracini, L 2019, 'Settler colonialism', in I Ness & Z Cope (eds), *The Palgrave encyclopedia of imperialism and anti-imperialism*, Springer Nature, pp. 1–5, https://doi.org/10.1007/978-3-319-91206-6_26-1.

Vidon, E, Rickly, J & Knudsen, D 2018, 'Wilderness state of mind: Expanding authenticity', *Annals of Tourism Research*, vol. 73, pp. 62–70, DOI: 10.1016/j.annals.2018.09.006.

von Scheve, C 2018, 'A social relational account of affect', *European Journal of Social Theory*, vol. 21, no. 1, pp. 39–59, DOI: 10.1177/1368431017690007.

Waitt, G, Figueroa, R & McGee, L 2007, 'Fissures in the rock: Rethinking pride and shame in the moral terrains of Uluṟu', *Transactions of the Institute of British Geographers*, vol. 32, no. 2, pp. 248–263, DOI: 10.1111/j.1475-5661.2007.00240.x.

Waldron, D & Newton, J 2012, 'Rethinking appropriation of the Indigenous: A critique of the romanticist approach', *Nova Religio*, vol. 16, no. 2, pp. 64–85, DOI: 10.1525/nr.2012.16.2.64.

Walker, K & Moscardo, G 2016, 'Moving beyond sense of place to care of place: The role of Indigenous values and interpretation in promoting transformative change in tourists' place images and personal values', *Journal of Sustainable Tourism*, vol. 24, no. 8–9, pp. 1243–1261, DOI: 10.1080/09669582.2016.1177064.

Walkerdine, V 2013, 'Using the work of Felix Guattari to understand space, place, social justice, and education', *Qualitative Inquiry*, vol. 19, no. 10, pp. 756–764, DOI: 10.1177/1077800413502934.

Wallis, G 2014, 'Transformative landscapes: Postcolonial representations of Uluṟu-Kata Tjuṯa and Tongariro National Parks', *Space and Culture*, vol. 17, no. 3, pp. 280–296, DOI: 10.1177/1206331213499470.

Walter, P 2021, 'Settler colonialism and the violent geographies of tourism in the California redwoods', *Tourism Geographies*, vol. 25, no. 1, pp. 243–264, DOI: 10.1080/14616688.2020.1867888.

Wang, N 1999, 'Rethinking authenticity in tourism experience', *Annals of Tourism Research*, vol. 26, no. 2, pp. 349–370, DOI: 10.1016/S0160-7383(98)00103-0.

Waterton, E 2010, *Politics, policy and the discourses of heritage in Britain*, Palgrave Macmillan, New York.

Waterton, E 2014, 'A more-than-representational understanding of heritage? The 'past' and the politics of affect', *Geography Compass, vol.* 8, no. 11, pp. 823–833, DOI: 10.1111/gec3.12182.

Waterton, E 2019, 'More-than-representational landscapes', in E Waterton & M Attha, (eds), *The Routledge companion to landscape studies, Routledge*, Abingdon, Oxon; New York, NY, pp. 91–101.

Waterton, E 2023, 'Performing identity and belonging at Pearl Harbor', *Geopolitics*, vol. 28, no. 4, pp. 1442–1464, DOI: 10.1080/14650045.2021.1901081.

Waterton, E, Burns, E & Willard, M 2016, *Photos of the past: The negotiation of identity and belonging at Australian tourism sites*, Western Sydney University, viewed 11 October 2022, https://researchdirect.westernsydney.edu.au/islandora/object/uws:38070/datastream/PDF/view.

Waterton, E & Dittmer, J 2014, 'The museum as assemblage: Bringing forth affect at the Australian War Memorial', *Museum Management and Curatorship*, vol. 29, no. 2, pp. 122–139, DOI: 10.1080/09647775.2014.888819.

Waterton, E, Smith, L & Campbell, G 2006, 'The utility of discourse analysis to heritage studies: The Burra charter and social inclusion', *International Journal of Heritage Studies*, vol. 12, no. 4, pp. 339–355, DOI: 10.1080/13527250600727000.

Waterton, E & Watson, S 2013, 'Framing theory: Towards a critical imagination in heritage studies', *International Journal of Heritage Studies*, vol. 19, no. 6, pp. 546–561, DOI: 10.1080/13527258.2013.779295.

Waterton, E & Watson, S 2014, *The semiotics of heritage tourism*, Channel View Publications, Bristol.

Waterton, E & Watson, *S* 2015, 'Methods in motion: Affecting heritage research', in B Knudsen & C Stage (eds), *Affective methodologies*, Palgrave Macmillan, New York, pp. 97–117.

Watkins, M 2016, 'Gauging the affective: Becoming attuned to its impact in education', in M Zembylas & P Schultz (eds), *Methodological advances in research on emotion and education*, Springer, New York, pp. 77–81.

Watkins, M, Noble, G & Driscoll, C 2015, *Cultural pedagogies and human conduct*, Routledge, London and New York.

Watson, S 2009, *Museums and their communities*, Routledge, New York.

Watson, S, Waterton, E, Smith, L, Smith, L, Waterton, E & Watson, S 2012, 'Moments, instances and experiences', in *The cultural moment in tourism*, Routledge, Abingdon, Oxon, pp. 1–16.

Wearing, S & Neil, J 1999, *Ecotourism: Impacts, potentials and possibilities*, Butterworth-Heineman, Oxford.

Wearing, S & Huyskens, M 2001, 'Moving on from joint management policy regimes in Australian national parks', *Current Issues in Tourism*, vol. 4, no. 2–4, pp. 182–209, DOI: 10.1080/13683500108667887.

Weiler, B & Black, R 2003, 'Nature, heritage and interpretive guide training - Mirror on the past window to the future', in R Black & B Weiler (eds), *Interpreting the land down under- Australian heritage interpretation and tour guiding*, Fulcrum Publishing, Colorado.

Weir, J, Woelfle-Erskine, C, Fuller, S, Diver, S & Higgins, M 2019, 'Investigating best practice: Doctoral fieldwork experiences with and without Indigenous communities in settler-colonial societies', *ACME an International e-Journal for Critical Geographies*, vol. 18, no. 6, pp. 1300–1322.

Wetherell, M 2013, 'Affect and discourse- What's the problem? From affect as excess to affective/discursive practice', *Subjectivity*, vol. 6, no. 4, pp. 349–368, DOI: 10.1057/sub.2013.13.

Wetherell, M, Smith, L & Campbell, G 2018, 'Introduction - Affective heritage practices', in L Smith, M Wetherell and G Campbell (eds), *Emotion, affective practices, and the past in the present*, Routledge, Oxon, pp. 1–22.

White, P 1957, *Voss*, Penguin Books, Harmondsworth.

Whitford, M & Ruhanen, L 2016, 'Indigenous tourism research, past and present: Where to from here?, *Journal of Sustainable Tourism*, vol. 24, no. 8–9, pp. 1080–1099, DOI: 10.1080/09669582.2016.1189925.

Whittington, V 2020, 'Bicentenaries and belonging: Public heritage of the lower Blue Mountains and the politics of place attachment', *Journal of Community Archaeology and Heritage*, vol. 7, no. 1, pp. 54–72, DOI: 10.1080/20518196.2019.1678812.

Whittington, V 2024, 'Memorialisation, reconciliation and truth-speaking: The role of explorer and massacre memorials in settler-colonial Australia', *Journal of Australian Studies*, vol. 48, no. 1, pp. 87–105, https://doi.org/10.1080/14443058.2023.2260397.

Whittington, V & Waterton, E 2021, 'Closing the climb: Refusal or reconciliation in Uluṟu-Kata Tjuṯa National Park?, *Settler Colonial Studies*, vol. 11, no. 4, pp. 553–572, DOI: 10.1080/2201473X.2021.2007749.

Whyte, K 2022, 'Against crisis epistemology', in B Hokowhitu, A Moreton-Robinson, L Tuhiwai Smith, C Anderson & S Larkin (eds), *Routledge handbook of critical Indigenous studies*, Routledge, New York and London, pp. 52–64.

Wijesinghe, S, Mura, P & Bouchon, F 2019, 'Tourism knowledge and neocolonialism - A systematic critical review of the literature', *Current Issues in Tourism*, vol. 22, no. 11, pp. 1263–1279, DOI: 10.1080/13683500.2017.1402871.

Williams, P & Chrisman, L 1994, *Colonial discourse and post-colonial theory: A reader*, Routledge, London, England; New York, New York.

Witcomb, A 2013, 'Understanding the role of affect in producing a critical pedagogy for history museums', *Museum Management and Curatorship*, vol. 28, no. 3, pp. 255–271, DOI: 10.1080/09647775.2013.807998.

Witcomb, A 2015a, 'Thinking about others through museums and heritage', in E Waterton & S Watson (eds), *The Palgrave handbook of contemporary heritage research*, Palgrave Macmillan, New York and Basingstoke, pp. 130–144.

Witcomb, A 2015b, 'Toward a pedagogy of feeling: Understanding how museums create a space for cross-cultural encounters', in A Witcomb and K Message (eds), *The international handbook of museum studies*, 1st ed., John Wiley & Sons, Hoboken, NJ, pp. 321–344.

Witcomb, A 2015c, 'Cultural pedagogies in the museum – Walking, listening and feeling' in M Watkins, G Noble and C Driscoll (eds), *Cultural pedagogies and human conduct*, Routledge, London and New York, pp. 158–170.

Witmore, C 2007, 'Symmetrical archaeology: Excerpts of a manifesto', *World Archaeology*, vol. 39, no. 4, pp. 546–562, DOI: 10.1080/00438240701679411.

Wodak, R 2001, 'What CDA is about – A summary of its history, important concepts and its developments', in *Methods of critical discourse analysis*, R Wodak and M Meyer (eds), Sage Publications, London, pp 1–13.

Wolfe, P 1999, *Settler colonialism and the transformation of anthropology*, Cassell, London and New York.

Wolfe, P 2006, 'Settler colonialism and the elimination of the native', *Journal of Genocide Research*, vol. 8, no. 4, pp. 387–409, DOI: 10.1080/14623520601056240.

Wright, S, Lloyd, K, Suchet-Pearson, S, Burarrwanga, L, Tofa, M & Bawaka Country 2012, 'Telling stories in, through and with Country: Engaging with Indigenous and more-than-human methodologies at Bawaka, NE Australia', *Journal of Cultural Geography*, vol. 29, no. 1, pp. 39–60, DOI: 10.1080/08873631.2012.646890.

Wu, Z & Hou, S 2015, 'Heritage and Discourse', in E Waterton & S Watson (eds), *The Palgrave handbook of contemporary heritage research*, Palgrave Macmillan, Hampshire, pp. 37–48.

Yuval-Davis, N 2006, 'Belonging and the politics of belonging', *Patterns of Prejudice*, vol. 40, no. 3, pp. 197–214, DOI: 10.1080/00313220600769331.

Zavala, M 2013, 'What do we mean by decolonizing research strategies? Lessons from decolonizing, Indigenous research projects in New Zealand and Latin America', *Decolonization: Indigeneity, Education & Society*, vol. 2, no. 1, pp. 55–71.

Zhang, H, Cho, T, Wang, H & Ge, Q 2018, 'The influence of cross-cultural awareness and tourist experience on authenticity, tourist satisfaction and acculturation in World Cultural Heritage Sites of Korea', *Sustainability*, vol. 10, no. 4, pp. 1–15, DOI: 10.3390/su10040927.

Index

Note: *Italicized* pages refer to figures and page numbers followed by "n" refer to notes.